AF327208

SYSTEMATICS OF THE GENUS COELOCNEMIS (COLEOPTERA: TENEBRIONIDAE)

A Quantitative Study of Variation

SYSTEMATICS OF THE GENUS COELOCNEMIS (COLEOPTERA: TENEBRIONIDAE)

A Quantitative Study of Variation

BY

JOHN T. DOYEN

UNIVERSITY OF CALIFORNIA PRESS
BERKELEY · LOS ANGELES · LONDON
1973

University of California Publications in Entomology
Advisory Editors: J. N. Belkin, R. M. Bohart, Paul DeBach, R. L. Doutt, D. D. Jensen,
W. H. Lange, Jr., E. I. Schlinger
Volume 73
Approved for publication September 22, 1972
Issued May 1, 1973

University of California Press
Berkeley and Los Angeles
California

◇

University of California Press, Ltd.
London, England

ISBN: 0-520-09481-6

Library of Congress Catalog Card Number: 72-619675

© 1973 by the Regents of the University of California

Printed in the United States of America

SYSTEMATICS OF THE GENUS COELOCNEMIS (COLEOPTERA: TENEBRIONIDAE)
A Quantitative Study of Variation

BY

JOHN T. DOYEN

INTRODUCTION

MEMBERS OF THE GENUS Coelocnemis are large, flightless beetles (pls. 1–3) which occupy conifer and oak-conifer woodland throughout most of North America west of the Rocky Mountains. In the Pacific Northwest their distribution is largely continuous, but elsewhere it is commonly disjunct, corresponding to the mosaic occurrence of appropriate habitats. In the southwestern deserts the genus is restricted to a few of the higher mountain ranges. Isolated populations also inhabit Santa Catalina and Cedros Islands as well as the mountains of extreme southern Baja California.

Since *Coelocnemis* has never been treated systematically, aside from isolated species descriptions, one purpose of this study is to provide a sound practical classification. To this end a formal taxonomic treatment of the genus is provided, including a key, descriptions, synonymies, and remarks concerning details of distribution and variation.

The broader intent of this investigation, however, is to detail the phenetic variation of the entire genus throughout its geographic range. The objective of this aspect of the study is to determine the nature of this variation and to correlate patterns of geographic variation with the probable evolutionary history of the genus. The fragmented nature of its distribution and the accompanying phenetic complexity make *Coelocnemis* an interesting subject for such an undertaking. Additionally, the geological and climatic history of western North America are known in sufficient detail to make it possible to determine the probable time of isolation of a number of the populations.

Several approaches are used to elucidate patterns of variation. Chromatographic analysis of defensive secretions provides one avenue of investigation. Larvae and some internal structures are examined, but most of the work consists of quantitative treatment of external morphological features of adults. Univariate comparisons among localities are used to illuminate geographic variation for each character. Multivariate analyses among localities include hierarchical (phenograms) and nonhierarchical ("taxometric maps") techniques. Diagrams called "phenomaps" are introduced as a simple means of portraying phenetic and geographic relationships simultaneously. These various methods are compared and evaluated and the results are integrated as far as possible to produce an interpretation of the phenetic relationships in the genus. Finally, the implications of this study to certain taxonomic concepts are discussed.

COLLECTING AND REARING PROCEDURE

More than 1,200 adults and about 25 larvae were collected in the field and maintained alive during the course of this study. Larvae are difficult to obtain in nature, but adults may be collected during any month of the year. Although existing collections contained very few records from winter and early spring it was found that this is a very profitable time to collect in areas which are not snow covered, since the adults aggregate during the cold months. Hibernating adults immediately resume feeding and reproduction at room temperature (about 21°C).

The beetles were maintained in rectangular, clear plastic boxes approximately 15 cm × 30 cm × 10 cm deep, and containing about 4 cm of substrate consisting of two parts commercial oakleaf litter and two parts peat. This mixture proved to be a suitable medium for larval development. A screened hole at least 10 cm × 5 cm in the container top is necessary to provide ventilation; excessive humidity causes high mortality in completely closed containers. Each container was furnished with slabs of bark or decomposing oak wood for oviposition and to allow adult concealment. Rolled barley, provided in weekly tablespoonful rations, proved to be a satisfactory diet, since it does not deteriorate so rapidly as wheat bran, and unlike whole grains the barley is soft enough to be edible. Cultures of this type must be watered every seven to ten days to maintain enough soil moisture for larval development. Both larvae and adults, however, can survive periods of several months with no free water. Each container, provisioned as described, is adequate for about 25 adult beetles.

Under the crowded conditions resulting from protracted oviposition, high larval mortality may occur because of cannibalism. This problem was avoided by isolating larvae about 2 cm in length in baby food jars with small holes punched in the lids. Individual larvae maintained in this manner survived to pupation and adult emergence with about 25 percent mortality.

TERMINOLOGY

Most specialized terms are explained when used. A few geographical terms which appear throughout the text are defined as follows: cismontane—west of the Sierra Nevada–Peninsular Range crest; transmontane—east of the Sierra Nevada–Peninsular Range crest; Peninsular Ranges—the mountain chains traversing a north-south axis in Northern Baja California and southern California; Transverse Ranges—the southern California mountain chains traversing an east-west axis, including the San Bernardino, San Gabriel, Tehachapi, San Rafael, and Santa Ynez Ranges.

ACKNOWLEDGMENTS

Many persons offered their help while this work was in progress. I am especially grateful to Howell V. Daly for guidance, encouragement, and many timely suggestions and criticisms during the entire course of this work. Kenneth Hagen and Lincoln Constance read the manuscript and made many valuable comments. Important criticism was also returned by Daniel Axelrod, John Chemsak, Wayne Moss, Constantine Slobodchikoff, and Perry Turner, Jr., who read all or part of the manuscript. Computational aspects of the work were greatly facilitated by consultation with Neil Bell and especially Wayne Moss. Walter Tschinkel graciously performed chemical analyses of defensive secretions and provided instruction in the use of the gas chromatograph and interpretation of its results. I am also indebted to James Haddock, Paul Opler, and Jerry Powell for contributing valuable insights to problems encountered during the research.

I offer my sincere thanks to the many persons who made the collections of various institutions available for study. These include: R. D. Anderson, College of Southern Utah, Cedar City; W. F. Barr, University of Idaho, Moscow; J. M. Campbell, Canadian National Collection, Ottawa; H. S. Dybas, Field Museum of Natural History, Chicago; G. F. Edmunds, University of Utah, Salt Lake City; S. Frommer and K. W. Brown, University of California at Riverside; W. J. Hanson, Utah State University, Logan; L. H. Herman, American Museum of Natural History, New York; M. T. James, Washington State University, Pullman; J. F. Lawrence and P. J. Darlington, Harvard University, Cambridge; H. B. Leech and P. H. Arnaud, California Academy of Sciences, San Francisco; J. A. Powell and J. Chemsak, University of California at Berkeley; R. A. Ring, University of Victoria, Victoria; P. O. Ritcher, Oregon State University, Corvallis; H. R. Roberts, Academy of Natural Sciences, Philadelphia; J. Schafrik, Cornell University, Ithaca; R. O. Schuster, University of California at Davis; H. Silfverberg Zoological Museum of Helsinki, Finland; T. J. Spilman, United States National Museum, Washington, D.C.; C. A. Triplehorn, Ohio State University, Columbus; F. S. Truxal and E. M. Fisher, Los Angeles County Museum of Natural History; F. G. Werner, University of Arizona, Tucson.

I am grateful to the following individuals who made available material from their personal collections or made special efforts to collect specimens for me: S. Ayala, K. W. Brown, J. M. Campbell, J. D. Haddock, K. Hagen, H. F. Howden, H. B. Leech, G. H. Nelson, P. A. Opler, J. A. Powell, G. Rotrammel, P. Schroeder, C. N. Slobodchikoff, R. Somerby, W. Tschinkel, C. A. Triplehorn, W. J. Turner, and D. Wood.

Wayne Moss wrote and provided copies of most of the FORTRAN IV programs (University of California Numerical Taxonomy Package, NTPAK) which were used in this work. Copies of the programs called UNIVAR and TAXMAP were obtained from D. M. Power, Royal Ontario Museum and University of Toronto, Toronto, and J. W. Carmichael, University of Alberta, Edmonton, respectively, to whom I am thankful. Finally, I acknowledge the great curatorial and secretarial assistance provided by my wife, Lisa, who also supplied the understanding and support necessary to complete this project.

Part of the research was carried out during the tenure of a National Science Foundation Traineeship. Statistical aspects of the work were done on a CDC 6400 computer using time provided by the University of California Computer Center.

BIOLOGY

Most comprehensive biological studies of North American Tenebrionidae concern economically important species, especially in the genera *Tenebrio, Tribolium,* and *Eleodes* (for example, Wade, 1921; Wade and St. George, 1923). The life history of *Bolitotherus cornutus* (Panzer) is described in detail by Pace (1967) and Liles (1956). The only published reference to the biology of *Coelocnemis* consists of fragmentary notes on the longevity and mating behavior of *C. magna* LeConte (Gissler, 1879a).

The following description applies primarily to the Arizona and California populations of *C. magna* and California populations of *C. californica* Mannerheim, which are most common in collections and are easily obtained in the field. Information-tion specific to Pacific Northwest and particularly Great Basin populations is mentioned where appropriate.

Habits of the Adults

Adults of both sexes of *Coelocnemis* are unusually long-lived among insects. Several hundred individuals have been maintained under laboratory conditions for over three years and three individuals of *C. californica* have survived for over five years. Longevity is also great in many other tenebrionid genera, including *Eleodes* (Gissler, 1879a), *Cratidus, Cryptoglossa, Nyctoporis* and *Phloeodes* (personal observation), *Blaps* (Tschinkel, personal communication), and *Bolitotherus cornutus* (Pace, 1967), suggesting that a long adult life is typical of the family. Many genera of the tribe Asidini are an exception, the adults surviving for only a few months.

Viable eggs have been obtained annually from many females of *Coelocnemis*, indicating that they remain reproductive throughout their lives, or at least for several years. The approximate age of an individual can be estimated by the degree of abrasion on the elytral and pronotal surfaces, which may be worn smooth, obliterating the original sculpturing in old individuals. Abrasion occurs as the beetles force their way through constrictions in decaying wood and beneath stones.

Adults of *Coelocnemis* overwinter in sheltered locations beneath the bark of tree snags, inside crevices in stumps, and infrequently beneath fallen logs or stones. Among the 200 or so individuals collected in California during the rainy months of December through March, fewer than 10 were taken from areas subject to wetting, indicating that this factor is critical in determining overwintering sites. Overwintering also entails marked aggregation of both sexes. Kaufman (1966) presents evidence that daily aggregation of *Blaps sulcata* Solier in favored shelters protecting them from heat is conditioned by the presence of traces of the beetles' defensive secretions, which act essentially as a pheromone in this case. Adults of the diverse genera *Cibdelis, Eleodes* (subgenera *Eleodes* and *Blapylis*), *Nyctoporis,* and *Phloeodes* frequently occur in mixed aggregations with one or more species of *Coelocnemis,* suggesting that the defensive secretions may function as a general attractant serving to assemble all these species in suitable locations for overwintering.

Aggregated masses of *Coelocnemis* typically number seven to 10 individuals, but up to 50 have been observed in a single overwintering site. Among eight collections of 12 or more overwintering individuals, 153 occurred in clusters of at least four, the average aggregation containing nine beetles. Only 18 specimens were taken in association with fewer than three others. Aggregation in *Coelocnemis* apparently occurs as a result of competition for the most favorable overwintering sites, possibly augmented by chemical attraction.

Adults collected during the cold season begin copulating and ovipositing within a few days under laboratory conditions, suggesting that overwintering involves only a state of quiescence rather than a condition of diapause.

During the months between March and September, depending on regional and local weather conditions, the overwintering aggregations break up and adults are found mostly in groups of two to four. Many types of cracks and crevices serve as daytime shelters during the warm season. At this time beetles are commonly found under fallen trunks and limbs, beneath debris scattered around snags, or even under stones and other horizontal or exposed surfaces which are very rarely utilized as overwintering sites.

The beetles normally remain hidden during the day throughout their period of activity, which lasts until October or December, depending on weather conditions. Frequently, at dawn or dusk or on cool, overcast days, individuals, sometimes in large numbers, wander from their shelters and may be seen feeding and copulating, but most activity is at night. Copulation and oviposition occur during spring and summer in nature, but copulation occurs continuously and oviposition sporadically throughout the year in laboratory populations.

LIFE HISTORY

Copulation occurs freely during the period of activity, the promiscuous adults pairing repeatedly and frequently. This tendency is pronounced enough under laboratory conditions that Gissler (1879a) noticed it, and is also evident under natural conditions, where copulating pairs of individuals are commonly collected. Study of marked individuals in laboratory populations demonstrated a high degree of variation in respect to frequency of copulation, with a minority of the individuals responsible for most of the pairings and other individuals not participating at all. This situation presumably applies to field populations as well, indicating that individual beetles may become reproductively inactive, at least temporarily.

Preceding copulation, the male mounts the female from the rear or side and clasps the forelegs around the prothorax anterior to the female's profemora or across the elytral humeri and clasps the posterior two pairs of legs around the abdomen. Mounted pairings may be maintained for periods greater than 12 hours. Although copulation occupies only a small fraction of this time, it may occur repeatedly during a single pairing. Impregnation is by means of an elongate, threadlike spermatophore, and requires several minutes for completion. Female specimens sometimes have the remains of an unsuccesfully introduced spermatophore protruding between the fifth sternite and elytra. A detailed account of spermatophore formation and transmission in *Pimelia* is given by Fiori (1954a, b). Production of fertile eggs by isolated females demonstrates that viable sperm may be stored by the female for periods of up to three months.

Oviposition has not been observed in the field, probably occurring at night. In the laboratory up to 50 eggs are usually deposited in small crevices in the substrate, especially in rotting wood. The unsculptured eggs are about 1.5 mm long by 0.75 mm broad, oval and white. About 14 days elapse before hatching. The larval life extends for one or sometimes two years, but the first stadium includes only one to three weeks of this period. The first instar larva is approximately 3 mm long, resembling the later instars except in the great relative length of the urogomphi and body setae, which both gradually decrease in relative size after each molt. Elongate setae may protect the delicate first instar larva from mechanical injury or predation by small insects or spiders; later instars develop a progressively tougher cuticle, obviating this function. After leaving the egg the larva immediately enters the substrate, either soft rotting wood or the soil, where the entire larval and pupal period takes place. As the larva grows, a hemicircle of six complexly wrinkled spines appears along the anterior edge of the tenth tergite, each developing from a thickening of the cuticle around a seta into large, multipointed structures in the mature larva. The urogomphi are apparently used in locomotion, the larva moving rapidly backward through its tunnels in rotten wood by using the urogomphi for traction. Touching the larva elicits a striking response with the terminal armature, suggesting a possible role in defense.

After five or six molts the larva reaches its mature size of approximately 5 to 6 cm, requiring approximately one year to pupation. The last instar larva excavates an irregular cell about 5 cm in diameter in the substrate in which it reclines on its side, the body curved ventrally, for 12 to 15 days before pupation. Pupation occurs between early May and early July in the laboratory and the limited number of pupae and prepupae collected in the field were also found within these dates. The pupal stage lasts 24 to 27 days in *C. magna,* and 19 to 24 days in *C. californica.* The newly emerged beetles remain teneral for two to three weeks, resting in the pupal chamber for this period, and apparently not reproducing until the following spring.

Habitat Preference

Few Tenebrionidae exercise rigid feeding requirements, but most display easily recognized habitat preferences. The species of *Coelocnemis* occur predominantly in woodland plant associations, seldom straying into other habitats. On the basis of habitat preference, three broad groupings of species may be recognized.

1) *Coelocnemis magna,* with morphologically distinct allopatric populations in cismontane California, montane Arizona, and certain of the desert mountain ranges and California islands, is a constant member of oak-woodland plant associations throughout most of its range. Exceptions occur in southern California and northern Baja California, where individuals are commonly taken in the belt of coastal chaparral, and in the desert mountain ranges, where they occupy pinyon-juniper woodland. The populations in the Owens Valley and associated basins represent the greatest habitat deviation of this group of populations. The vegetation here is transitional between that of the Great Basin and Mojave Desert, with pinyon-juniper woodland present only at elevations above about 1,500 m. Individuals of *C. magna* have been taken on the floor of the treeless Panamint Valley at less than 500 m elevation and in the Owens and Coso valleys at elevations less than 1,000 m, and only a minority of known records are from localities higher than 1,500 m.

In cismontane California and montane Arizona, where its distribution is relatively continuous, *C. magna* is a common element of the oak-woodland habitat, displaying a marked preference for the wood of dead oaks for sheltering and overwintering sites. It is important to emphasize that this preference includes all the treelike oak species in a given area. In California material has been collected from snags of *Quercus agrifolia* Nee, *Q. lobata* Nee, *Q. wislizenii* De Candolle, *Q. chrysolepis* Liebmann, *Q. kelloggii* Newberry, *Q. douglasii* Hooker &

Arnott, and *Q. garryanna* Douglas. Only a few collection records from *Populus* sp., *Platanus* sp., and *Pinus coulteri* D. Don and a single record from *P. sabiniana* Douglas are known, markedly contrasting with the preference for oaks. Tree associations of the Arizona populations include *Q. arizonica* Sargent, *Q. gambellii* Nuttall, *Q. reticulata* Humboldt & Bonpland, *Q. hypoleucoides* A. Camus, *Q. undulata* Torrey, and isolated records from *Prunus* sp., *Pinus ponderosa* Douglas, and *Pseudotseuga menzesii* (Mirb.) Franco.

2) The populations considered under *C. sulcata* Casey occur in drier areas of the Arizona oak-woodland and in the Great Basin. The Arizona populations of *C. sulcata* occupy habitats very similar to those of *C. magna,* while the Great Basin populations of *C. sulcata,* together with the populations considered under *C. punctata* LeConte, occur sympatrically in pinyon-juniper woodland habitats or occasionally in dry, open *Pinus ponderosa* forests. The distribution of both *C. sulcata* and *C. punctata* is highly disjunct throughout most of their ranges, due to the basin-and-range topography of the Great Basin. Limited intercommunication between populations isolated in the various mountain ranges is indicated by scattered records from intervening basins. Neither of these Great Basin species exhibits the strong preference for standing dead wood shown by *C. magna,* probably because of the tendency of the bark of pinyon and juniper to remain tightly adherent to the dead trunks and because of the open nature of the woodland which exposes dead trees to full insolation. Consequently *C. sulcata* and *C. punctata* are most frequently discovered sheltering beneath stones and piles of rubble on canyon sides and bottoms and are commonly found in the course of ground searches at dawn or dusk and on overcast days.

3) The diverse populations considered under *C. californica, C. lucia* n. sp., and *C. rugulosa* n. sp. all inhabit transition zone oak-conifer woodland or Canadian zone conifer woodland at elevations from near sea level to over 3,000 m. *Coelocnemis californica* apparently requires a more mesic habitat than the species discussed above, but occurs in partial sympatry with *C. magna* in the upper transition oak-woodland in northern California and in complete sympatry in the mountains of southern California. *Coelocnemis magna* is also sympatric with *C. lucia* in the Santa Lucia Mountains in Monterey County and *C. californica* with *C. rugulosa* in northern California. Where sympatry occurs the same trees often harbor both species.

Whereas individuals of *C. magna* show a clear preference for dead oak wood as sheltering sites, those of *C. californica* and *C. lucia* may be associated with either dead oaks or conifers, depending on local woodland composition. Recorded hosts in California include *Q. kelloggii, Q. chrysolepis, Q. agrifolia, P. ponderosa, P. coulteri, Pseudotsuga menzesii, P. macrocarpa* (Vasey) Mayr, and *Sequoia sempervirens* (D. Don) Endlicher. Oaks appear to be preferred where present and are probably the only trees utilized in the Alameda and Contra Costa County portion of the range where the native conifers consist of closed-cone pines, not known to be used as shelters, and small, isolated stands of redwood. Similarly, the disjunct populations in the Sierra Madre in Santa Barbara County occur in moist canyon bottoms dominated by *Quercus chrysolepis,* with *Pinus coulteri* limited mainly to ridge tops. These are the only two substantiated instances in which *C.*

californica occurs apart from conifers. Throughout much of its range in Oregon, Washington, Idaho, and British Columbia, oaks are absent and, although the only recorded host is *Pseudotsuga menzesii,* other conifers undoubtedly serve as sheltering and overwintering sites.

As in the case of *C. magna,* the suitability of various trees species is probably dependent on the characteristics of bark dehiscence. Bark which curls away in large slabs or plates creates suitable spaces for entry and concealment, while tightly adherent bark, such as that of *Abies,* or bark which cracks into numerous small plates, such as that of *Arbutus,* fails in these requirements. The preference for dead oak wood may be explained by its resistance to weathering, enabling it to provide a hard substrate to which the beetles are able to cling. Extremely rotten, disintegrating timbers are rarely chosen for sheltering or overwintering.

Neither *C. slevini* Blaisdell nor the undescribed species from the cape region of Baja California have been collected by the author, and nothing is known of their ecological requirements.

FEEDING HABITS

Although a substantial portion of the Tenebrionidae restrict their feeding to fungi of various sorts (e.g., *Diaperini*), the food of most includes a variety of dead plant and occasionally animal material. Recorded observations of feeding in the field are infrequent, but laboratory rearing of at least one complete generation of more than 30 tenebrionid species in six tribes has been accomplished by the author by using commercial grains such as rolled barley as both larval and adult diet, suggesting generalized nutritional requirements. *Coelocnemis* differs from many other tenebrionids reared by the author in that it requires particulate matter in the adult diet. Whole grains, various fungi, lichens, and fresh vegetable material are accepted, while disintegrating grain products are refused. Rotting wood or humus is ingested by the larvae, but whole grains and other particulate vegetable matter is also accepted.

Eugregarine protozoans are common commensals in the midgut of adult and larval *Coelocnemis.* A large number of tenebrionid gregarines have been described by Theodorides (some selected references are: Theodorides, 1952, 1966, and Theodorides and Jolivet, 1964) and are probably universally present in the family. The number of commensals in an individual of *C. magna* varies from none to more than 100. In *Tenebrio,* gregarines are not necessary for normal growth and development, but their absence significantly reduces final pupal weight and pupation success of larvae reared on suboptimal diets (Harry, 1969). Possibly this may be an important factor influencing the appreciable variation in adult size of *Coelocnemis* and many other tenebrionids.

DEFENSIVE SECRETIONS

Members of many families of Coleoptera synthesize chemicals which are used in defense, the Carabidae, Staphylinidae, and Tenebrionidae being notable examples. The first mention of repugnatorial glands in tenebrionids is that of Gissler (1879*b*; *Eleodes*). The accounts of Eisner and Meinwald (1966), Happ (1968), Ladisch (1967), Roth and Eisner (1962), and Schildknecht et al. (1964) will provide an entry to the recent literature. The general subject of chemical interactions between species is reviewed by Whittaker and Feeny (1971). A comprehensive account of the gross anatomy of the defensive glands in the related families Alleculidae and Lagriidae is given by Kendall (1968). Various workers have suggested that the glandular morphology or the composition of coleopterous defensive secretions might be of taxonomic significance, but the only comprehensive documentations are those of Moore and Wallbank (1968) in carabids and Tschinkel (in progress) in tenebrionids. Both these studies emphasize differences at the generic and higher levels of classification. The results discussed here are primarily intended to explore the value of defensive secretion composition at the specific and infraspecific levels.

Among the Tenebrionidae, only the members of the subfamily Tenebrioninae utilize chemical secretions as a means of defense, suggesting, with the correlated differences in morphology, that the Tenebrionidae, as presently defined, may be a polyphyletic family. The subfamily Tenebrioninae, whose glands are apparently homologous to those in the Alleculidae and Lagriidae (Kendall, 1968), is probably more closely related to those families than to the tenebrionid subfamilies Asidinae and Tentyriinae (Doyen, 1972).

The defensive glands in *Coelocnemis* consist of diffuse glandular tissue surrounding large, paired, distensible reservoirs opening posterad of the seventh abdominal sternite. Their gross structure is similar to that of the glands of *Eleodes longicollis* LeConte (Eisner et al., 1964) and *Cibdelis blaschkei* Mannerheim (Doyen, unpublished). The defensive secretion is voided as a liquid which spreads anteriorly over the elytra and abdominal sterna, and is never sprayed, as in *Eleodes* (subgenus *Eleodes*) and in *Scotobaenus parallelus* LeConte. *Coelocnemis* practices the common tenebrionid habit of posturing with the head lowered and abdomen elevated during and after release of the secretion. A given individual never releases its entire reserve of defensive material after a single stimulation, but is capable of repeated ejections. Habituation occurs rapidly in captivity, considerable stimulation being required to elicit release by individuals that are handled frequently. Although the predators of *Coelocnemis* are unknown, it is probable that rodents are important, as Eisner et al. (1964) have shown for *Eleodes longicollis* LeConte.

Secretions were collected from living beetles into clean test tubes, extracted with carbon disulfide, dried with anhydrous magnesium sulfate, and stored in closed vials at 0°C or lower. Gas-liquid chromatographic analysis was done using a Varian Aerograph Model 1200-1 with a flame ionization detector, on 5% SE-30 at 75°C. Temperature programming was linear at 10°C per minute. The results of each chromatograph represent the pooled secretions of as many individuals as possible (up to 20) from each geographic locality. A total of more than 30 samples was analyzed, including several replicates each of *C. magna* LeConte, *C. sulcata* Casey, *C. californica* Mannerheim, *C. lucia* n. sp., and *C. punctata* LeConte.

The composition of the defensive secretions clearly separates *Coelocnemis* from other related genera, but does not show differences sufficient to distinguish species. Representative chromatographs of *Coelocnemis* species analyzed show that the secretions contain aproximately 20% p-toluquinone, 50% p-ethylquinone, and 30% 1-tridecene,[1] represented by the three large peaks labeled A, B, and C, respectively, in figs. 1–4. The balance consists of small amounts of 1-undecene, appearing as a shoulder on the p-ethylquinone peak, 1-pentadecene and several unidentified substances (peaks labeled a, b, c, d, e, f). Consistent species differences involve only the minor components, particularly the peak labeled (b) and the series of peaks at 145 to 165°C. On the basis of these minor components *Coelocnemis* may be subdivided into one species group containing *C. californica* Mannerheim, *C. rugulosa* n. sp., *C. lucia* n. sp., and *C. punctata* LeConte, and another containing *C. magna* LeConte and *C. sulcata* Casey. This subdivision of the genus is also apparent on morphological, ecological, and zoogeographic grounds, as will be shown later. Consequently, the species groups designated Californica and Magna are recognized and referred to throughout this paper. A detailed discussion of the taxonomic composition of these groups and the evidence for them is presented in subsequent sections of this paper.

Separate chromatographs of pooled secretions of males and females and of individuals disclosed appreciable variation in the relative proportions of the major components and in the presence of minor components, but no chromatograph was so variable that it could not be assigned to the proper species group. Extremes of observed variation among individuals of one collection are shown in figures 3–4. Secretions of selected groups of individuals chromatographed shortly after collection and after periods of 6 and 12 months in captivity revealed no significant differences in composition.

Within the two species groups designated above only minor and unreliable differences in defensive secretion composition appear in the chromatographs. In the Californica group (fig. 1) variation between populations of a single species is as pronounced as that between species and the only differences involve the proportions of various components. In the Magna group, the Great Basin and montane Arizona populations (*magna* and *sulcata*) uniformly show three minor peaks of approximately equal amplitude eluting at 145 to 165°C (fig. 2). The corresponding portion of the chromatographs of California populations of *magna* (figs. 3–4) shows a more variable series of peaks, indicating the presence of one to five separate components, but frequently there are three peaks represented, apparently corresponding to those of the Great Basin and Arizona populations. Since the Magna species group has evolved three species and many infraspecific entities showing appreciable morphological differences, it is obvious that the evolutionary rate of change of defensive secretions has been considerably more conservative than that of morphological change. A similar trend is evident in the Californica species group, and also occurs in the related genera *Iphthimus* and *Cibdelis,* indicating that defensive secretion differences in the Tenebrionidae may be most valuable at the generic or higher categories. However, in large genera such as *Eleodes* much greater diversity exists in defensive secretion composition (Tschinkel, 1968) and gas-liquid chromatography promises to clarify relationships in such cases.

---

[1] Chemical identifications performed by Walter Tschinkel, Florida State University.

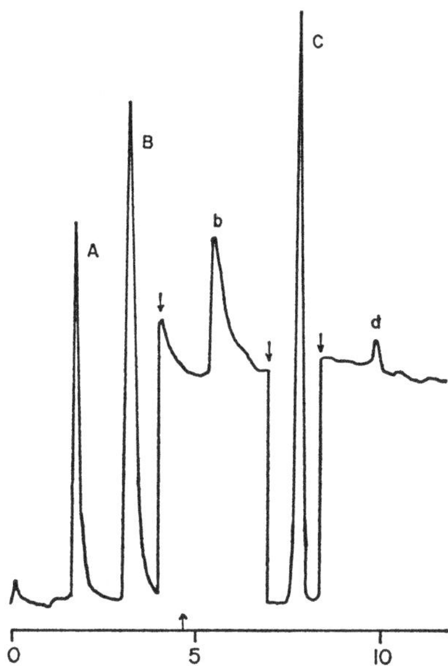

Fig. 1. Chromatograph of pooled defensive secretions of 18 individuals of *Coelocnemis punctata* LeConte from Mono County, California, showing composition typical of the Californica species group. Upward pointing arrow indicates onset of temperature program. Downward pointing arrows indicate changes in attenuation during the course of the analysis. Scale represents time in minutes.

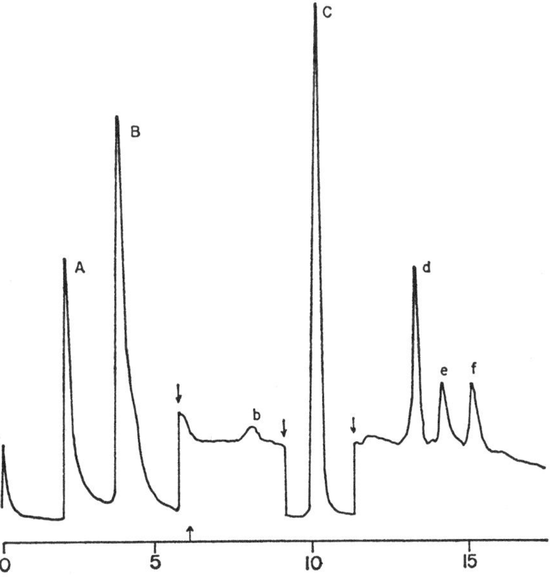

Fig. 2. Chromatograph of pooled defensive secretions of 8 individuals of *Coelocnemis magna* LeConte from Greenlee County, Arizona. This composition is also typical of *C. sulcata* Casey from throughout its range.

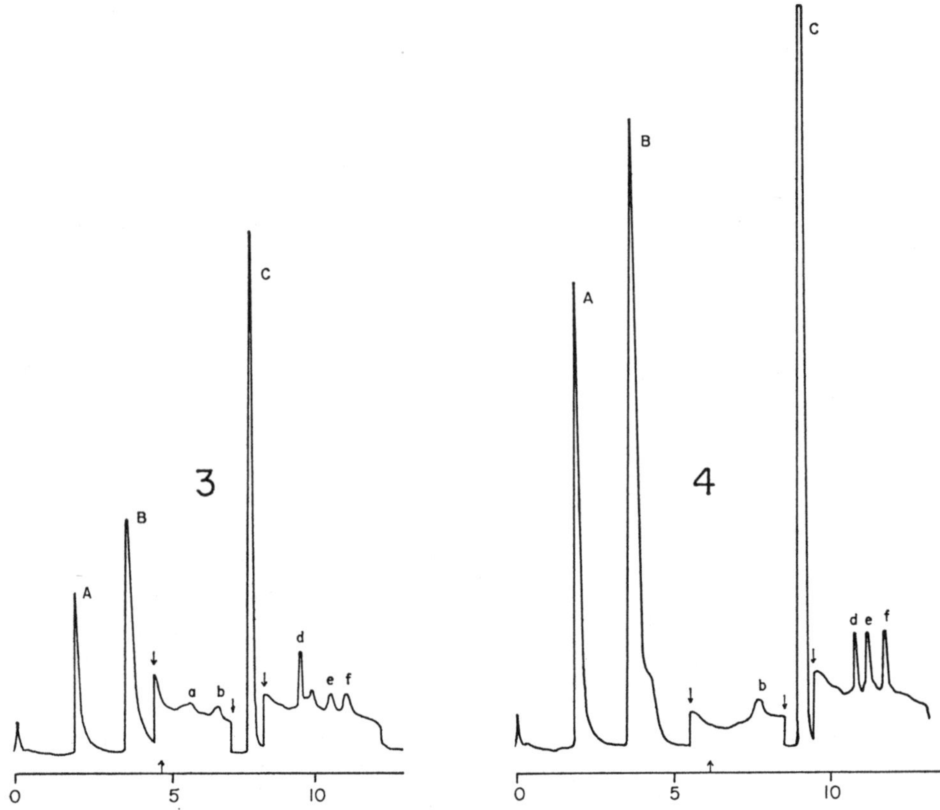

Figs. 3–4. Chromatographs of two individuals of *Coelocnemis magna* from Orange County, California, illustrating normal range of individual variation encountered within populations.

The secretions of *Iphthimus lewisi* Horn, *Scotobaenus parallelus* LeConte, *Cibdelis blaschkei* Mannerheim,[2] *Alobates pennsylvanicus* Degeer, and *Oenopion zopheroides* (Horn), all members of the tribes Tenebrionini and Coelometopini, were also analyzed in order to clarify the generic position of *Coelocnemis.* Among the species examined, *Scotobaenus parallelus* appears to be most similar to *Coelocnemis* regarding defensive secretion composition. Its chromatograph (fig. 5) reveals major peaks exactly homologous to those of *Coelocnemis,* but the identity of these peaks has not been verified by other techniques. The morphological peculiarities of *Scotobaenus,* discussed in the systematics section below, and the fact that its secretion is sprayed, preclude the possibility of close relationship to *Coelocnemis.* As shown later, *Coelocnemis* is morphologically quite similar to the genus *Iphthimus,* whose chromatograph is shown in figure 6. The chromatographs of these two genera differ in the absence of 1-tridecene and other components of high volatility in *Iphthimus,* and in the relative amounts of the shared components. The chromatograph of *Alobates pennsylvanica* (fig. 7) is very similar to that of *Iphthimus,* although these two are rather different morphologically. *Oenopion*

---

[2] Chromatograph for this species taken from Tschinkel (1968)

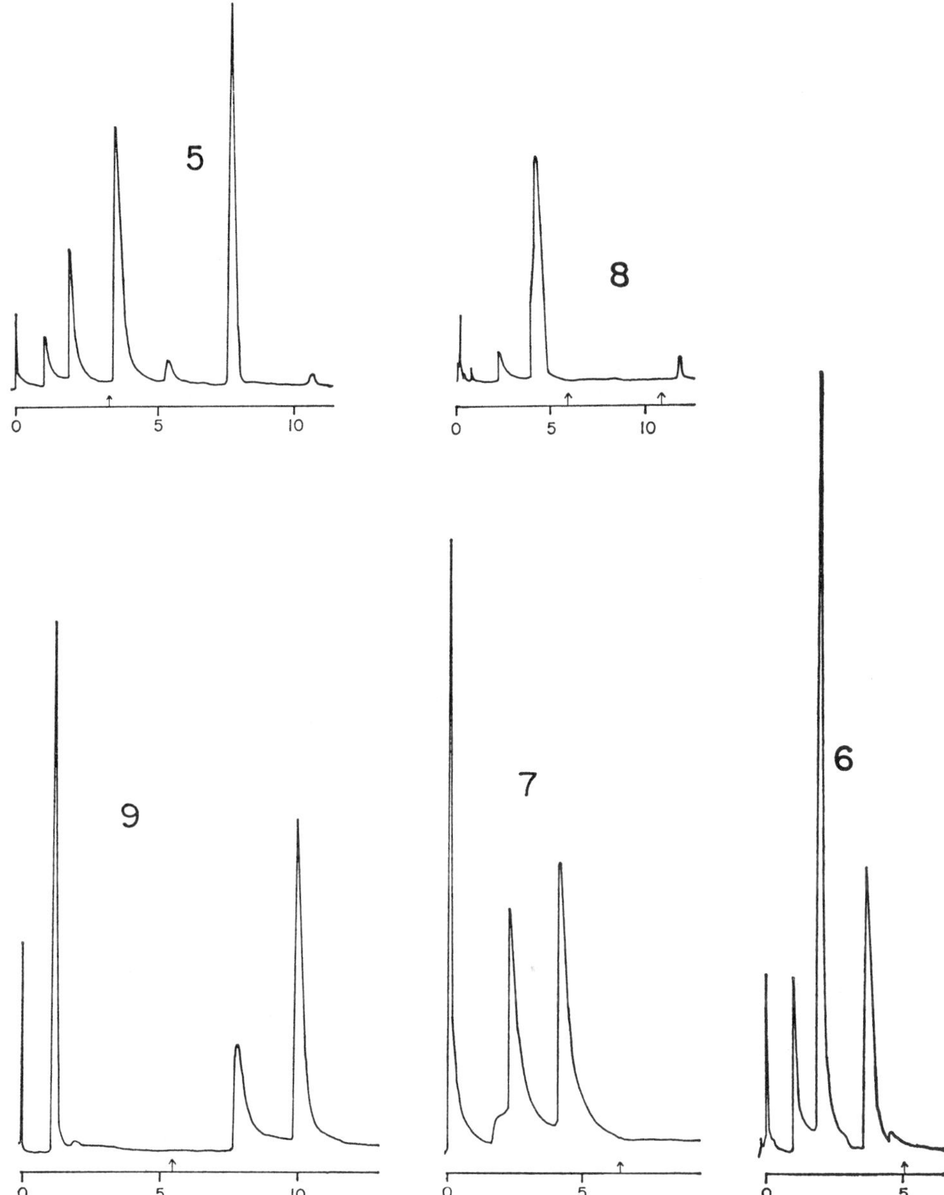

Figs. 5–9. Chromatographs of pooled defensive secretions of selected species of tenebrionine tenebrionids. Fig. 5. *Scotobaenus parallelus* LeConte (4 individuals). Fig. 6. *Iphthimus lewisi* Horn (15 individuals). Fig. 7. *Alobates pennsylvanicus* Degeer (5 individuals). Fig. 8. *Cibelis blaschkei* Mannerheim (30 individuals). Fig. 9. *Oenopion zopheroides* Horn (3 individuals).

zopheroides, formerly included in *Iphthimus,* shows little similarity to other species of that genus either in defensive secretions (fig. 9) or morphology (Doyen, 1971). *Cibdelis* is isolated among genera discussed here both chemically (fig. 8) and morphologically.

The most useful conclusion indicated by the results presented here is that closely related species share a common composition and method of delivery of defensive secretions. The limited number of genera studied possess characteristic secretions, but generic relationships suggested by comparison of chromatographs are not always concordant with those indicated by morphological similarity (e.g., *Iphthimus* and *Alobates*). Possibly, the limited number of components occurring in tenebrionid secretions creates a mosaic of patterns, with convergence at the generic level. Such mosaic patterns occur in the Carabidae, where species of distantly related genera secrete compounds such as formic acid and methacrylic acid (Moore and Wallbank, 1968). Below the generic level, at least in *Coelocnemis*, defensive secretions are nearly invariant, and of little value in taxonomic discrimination. Uniformity of defensive secretion composition also occurs within genera of Carabidae (Moore and Wallbank, 1968), where the only differences are trivial changes in the proportions of substances present.

In both families, defensive secretions will probably be most valuable for taxonomic purposes when unusual compounds or methods of release are present. For example, in the Tenebrionidae, Nearctic members of the tribe Scaurini are characterized by secretions containing unusual components (including naphthoquinones) not found in other tenebrionids, and by eversible reservoirs (Tschinkel, 1972). In *Eleodes* (subgenus *Eleodes*), the composition of secretions is not remarkable (mostly a mixture of quinones), but all species eject the secretion as fine jets of spray which may travel several feet. In both examples, the chemical evidence supports classifications based on morphological similiarities.

BIOMETRICS
MATERIALS

The specimens used in analyzing morphological variation were selected primarily from museum material; specimens field-collected during the study were added to increase certain sample sizes and to fill geographic gaps in the material at hand. Nevertheless, although more than 6,000 museum specimens were examined and more than 1,200 additional specimens were collected by me in 1968 and 1969, most of the samples are quite small by statistical standards because of the large geographic area encompassed. Many of the sample sizes could have been increased by enlarging the areas included in each locality, but it was deemed equally important to limit localities to a reasonably small and ecologically uniform region. The resulting samples and localities are a compromise dictated by these two requirements. The desire to keep the total number of individuals included at a manageable level imposed an overall limit on sample size; consequently, except in a few cases no sample contains more than 15 specimens.

The final choice of specimens was based on their morphological condition and completeness and presumed veracity of label information. Whenever possible a sample includes specimens collected at the same time and locality, but as indicated

---

Fig. 10. Distribution of sample localities. The alphabetic designations match those listed in the appendix. Closed circles indicate Californica species group samples. Open circles indicate Magna species group samples. The maps appearing here and in other portions of the text are intended to show the major topographic features of western North America in a simplified

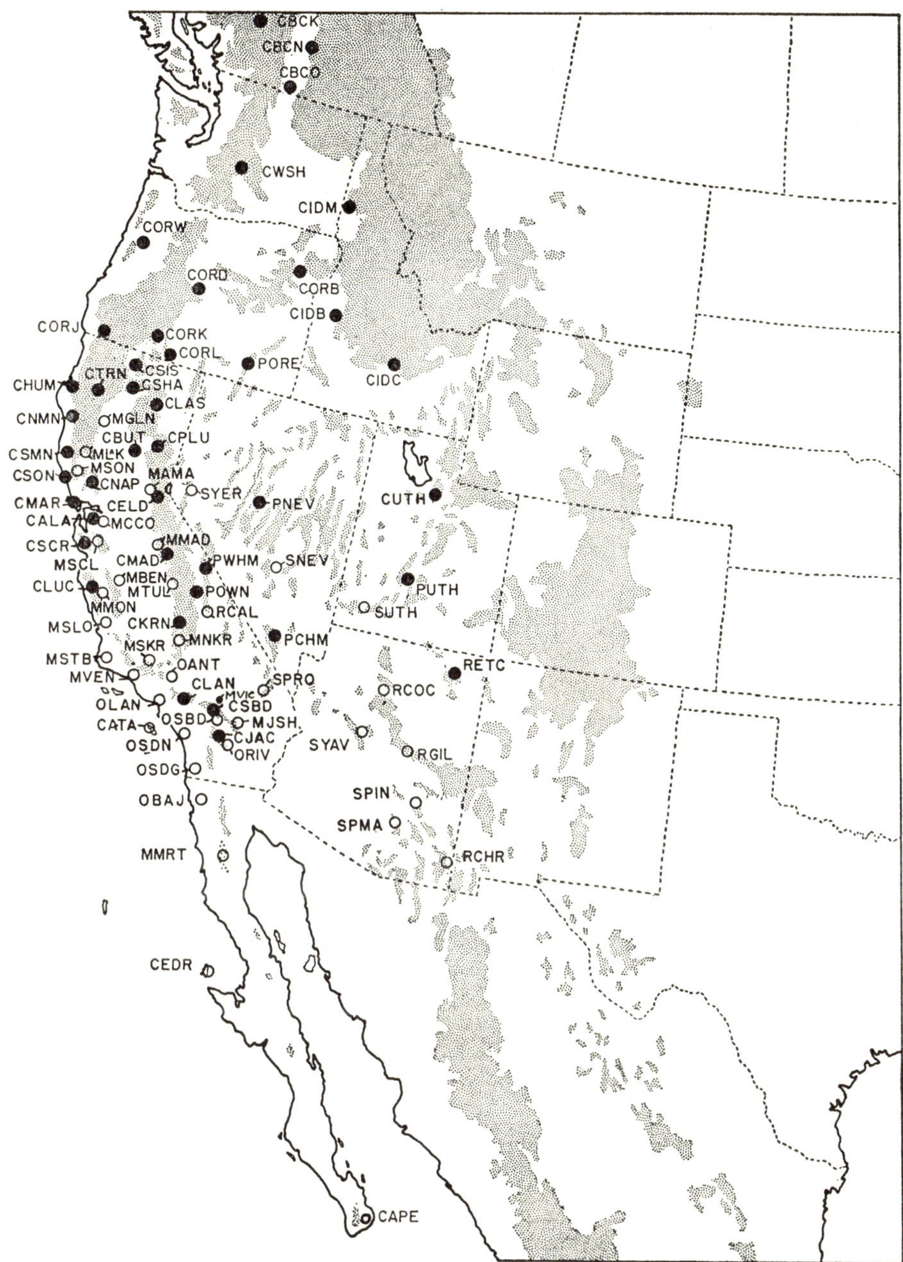

fashion. They were constructed by modifying contour and vegetation maps. Throughout most of the included area a 5,000 ft. contour was chosen to delimit important mountain ranges. Exceptions occur in central coastal California (Lat. 35° to 38°) and in the Basin and Range Province. In the former region the dominant mountain regions are relatively low, so that a 2,000 ft. contour was selected to reveal these ranges, which exert a profound influence on local vegetational patterns. The numerous ranges of the Basin and Range Province arise from an elevated plateau which is frequently higher than 5,000 ft. For this situation a 7,000 ft. contour was utilized. The resulting map matches fairly closely the distribution of coniferous woodland in western North America.

above most are composites. A total of 81 localities from throughout the range of
the genus was chosen for detailed examination and sexes were analyzed separately.
In the text and on illustrations, samples are referred to by 4-character alphabetic
designations intended to suggest the sample locality. The sample abbreviations and

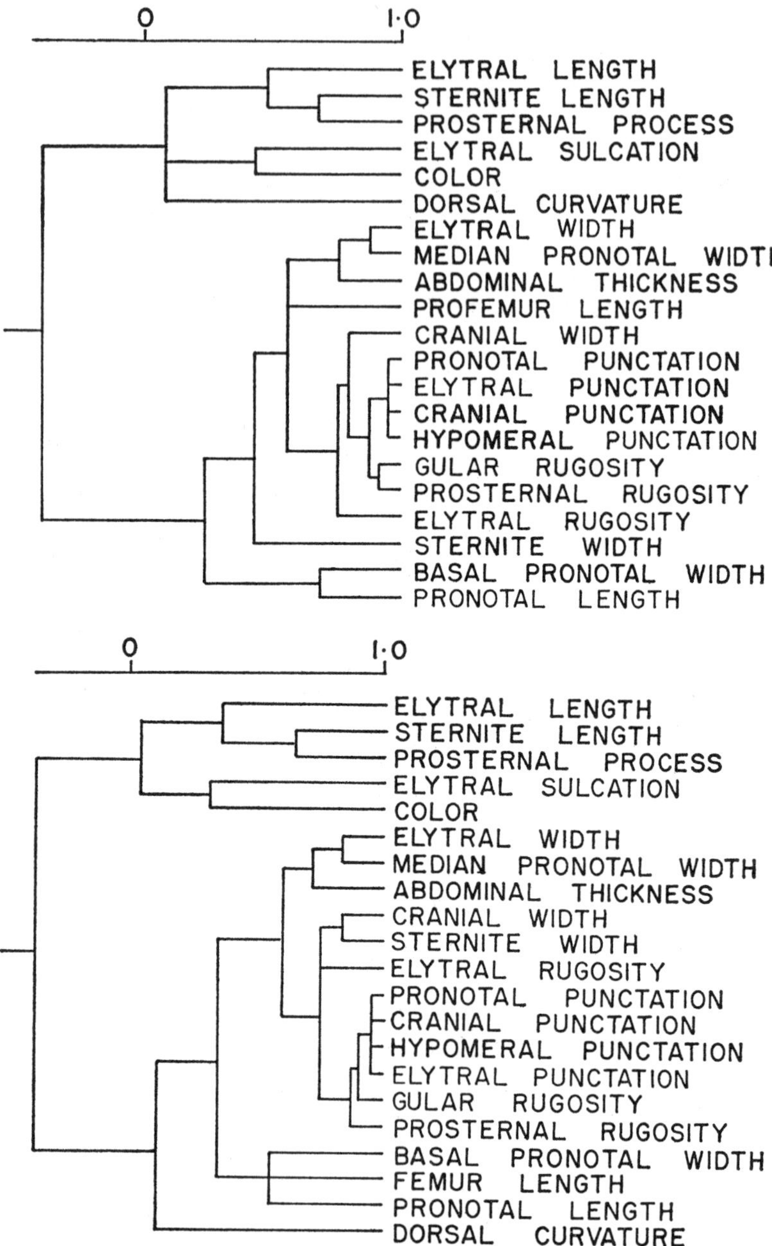

Fig. 11. Correlation phenograms of unstandardized characters clustered by
unweighted pair group method. Females above, males below.

sizes and geographic composition of each locality are categorized according to species in Appendix 1. The exact composition of the species groups mentioned in the section on defensive secretions is also specified in Appendix 1. These species groups are referred to extensively in subsequent discussions of variation. The approximate location of each sample locality is shown in figure 10.

CHARACTERS

A preliminary study of variation involving about 200 specimens was performed to select the characters used in the detailed analyses. Only characters showing significant variation between at least two localities were chosen for final analysis, but difficulty of measurement eliminated several features which otherwise were suitable. The final roster of characters includes most of the morphological features which were originally used to define the described species of the genus. The 21 characters eventually selected are defined in table 1.

Correlations among the morphological characters were examined by hierarchical cluster analysis of Pearson's product-moment correlation coefficients, treating characters as OTUs (= operational taxonomic units; Sokal and Sneath, 1963), and localities as characters. For this analysis the data matrix was not standardized. The phenograms (UPGA) (= unweighted pair group method using average linkage; Sokal and Sneath, 1963) of male and female characters each break into two distinct clusters differing chiefly in the placement of *curvature of dorsum* (fig. 11). This is primarily a secondary sexual character, males of all populations showing greater ventral curvature of the posterior abdomen than the females, but Sierra Nevadan populations of *C. californica* show definite flattening of the elytral dorsum in both sexes.

In both male and female phenograms the four punctation characters are correlated at a very high level, suggesting that all but one could have been deleted. Two of the three rugosity characters are also rather tightly linked with this punctation cluster. These characters describing cuticular sculpturing show consistent differences between the two species groups composing the genus, but their adaptive significance is unknown. It is of interest that rugosity and punctation characters have been used to distinguish several of the accepted species of *Coelocnemis*. The high correlation among these characters suggests that certain species differ in very few independent characters and may not warrant taxonomic recognition.

The remainder of the characters fall into rather diffuse clusters with little suggestion of redundancy. *Femur length,* a secondary sexual character, and *fifth-sternite width* differ between the sexes in their position in the large cluster containing most of the characters. Other characters show very similar linkage relationships in both sexes. The cophenetic correlation coefficients are 0.921 for both males and females.

INDIVIDUAL CHARACTER VARIATION (UNIVARIATE ANALYSIS)

The basic sample statistics (mean, standard deviation, etc., not reproduced here) were computed for each locality sample for use in assessment of variation in single characters and of relationships of overall phenetic similarity. Analysis of variance over all localities in each species group was performed for every character using

TABLE 1

Summary of characters used in numerical analyses. Linear measurements were made with dial calipers. *Elytral length, femur length,* and dimensions of the fifth sternite were measured to the nearest 0.05 mm, all other dimensions to the nearest 0.01 mm. The raw measurements of elytral length were used in all analyses. Other measurements were converted to ratios to elytral length to reduce the influence of size in estimates of phenetic relationships. Values for rugosity, punctuation, shape of the prosternal process and color were coded by comparison with reference series of specimens. Unmodified values for qualitative (coded) characters were used in all analyses.

Pronotal length: The distance along the midline between the anterior and posterior pronotal margins.

Median pronotal width: the greatest transverse distance across the pronotum, usually near the pronotal midpoint.

Basal pronotal width: the transverse distance between the posterior pronotal angles.

Elytral length: the shortest distance from the front margin of the scutellum to the elytral apex.

Elytral width: the greatest transverse distance through the elytra, typically near the elytral midpoint.

Abdominal thickness: the median dorsoventral distance between the elytral dorsum and the venter at the approximate midpoint of the first abdominal sternite.

Profemur length: the distance from the proximal margin of the trochanter to the apex of the femur, measured along the posterior femoral margin.

Cranial width: the geratest transverse dimension of the head, measured across the frons and genae just anterior to the eyes.

Fifth sternite length: the midline distance from the posterior margin of the membrane separating ing visible sternites four and five to the posterior margin of sternite five.

Fifth sternite width: the width of the fifth sternite measured across its anterior corners.

Elytral rugosity: the rugosity of the elytral disk at approximately the longitudinal midpoint. Coded from 1 (smooth) to 6 (coarsely rugose).

Postoral rugosity: the rugosity of the region just posterad of the eyes and insertion of the mouthparts. Coded from 1 (slightly rugose) to 3 (coarsely rugose).

Prosternal rugosity: the rugosity of the anteromedial cuticle of the prosternum. Coded from 1 (slightly rugose) to 3 (coarsely rugose).

Elytral punctation: the punctation of the elytral disk at the approximate longitudinal midpoint. Coded from 1 to 3 according to increasing coarseness of punctation.

Pronotal punctation: punctation of the central region of the pronotal disk. Coded from 1 to 3.

Cranial punctation: punctation of the central part of the vertex between the eyes. Coded from 1 to 3.

Punctation of hypomeron: punctation of the deflexed portion of the pronotum, recorded from the area midway between the coxal cavity and the lateral pronotal margin, coded from 1 to 3.

Prosternal process: the prominence of the prosternal process. Coded from 1 (declivous, flattened, rounded posteriorly) to 3 (prominent, horizontal, acutely angulate posteriorly).

Curvature of dorsum: the curvature of the elytral dorsum in lateral aspect. Coded from 1 (relatively flat) to 3 (relatively convex).

Elytral sulcation: The degree of sulcation of the elytral disk at the approximate midpoint. Coded from 1 (planar) to 3 (distinctly sulcate).

Color: the color of the elytral disk. Coded from 1 (black) to 2 (deep reddish black).

Power's (1970) version of Gabriel's (1964) simultaneous test procedure (STP), a multiple comparison test for differences between means. This technique designates maximal homogeneous subsets in a list of locality means ranked in decreasing order of magnitude. Subsets may be overlapping as they are in this study, with the exception of only a single character (color) which is invariant in all but a few samples. Subsets are marked by vertical lines beside the lists of ranked means; the probability level is 0.05 in all cases. The shaded circles appearing on the distribution map and to the left of the list of means do not coincide with the subsets just described but to a second, independent grouping of means determined by dividing the total range of variation for each character into five equal intervals. In each case the largest values are represented by closed circles, the smallest by open circles. It should be noted that the divisions marked by the shaded circles do not indicate significant character differences for those characters in which the maximal homogeneous subsets are broadly overlapping.

The Magna and Californica species groups, which were defined on the basis of biochemical and ecological characteristics (p. 10) are considered separately in the analyses of variation in single characters. The morphological integrity of these species groups is clarified in the discussion of overall morphological variation (pp. 43–49, figs. 38–43). Separate STP diagrams were constructed for the two sexes for each species group (Magna and Californica), males and females showing different ranges of means for several characters. Geographic patterns of variation are typically very similar in each sex, and only the results for females are illustrated unless obvious differences appear in the males. Complete analysis of male characters is presented in Doyen (1970). Some localities (asterisked in Appendix 1) which appear in the phenograms presented later are not in these diagrams because of the extremely small samples available. Only one sex of some samples is represented for the same reason. More precise identification of sample localities than is possible from the STP maps may be made by comparison with figure 10 and Appendix 1. The implications and interpretation of STP results are discussed in detail by Gabriel and Sokal (1969) and Power (1970).

RESULTS OF UNIVARIATE ANALYSIS

Pronotal length.—Relative pronotal length varies in a different manner in the two species groups. In the Magna group (fig. 12) the Great Basin and Arizona populations of both *magna* and *sulcata,* the transmontane California populations of *magna* and the Cedros Island species, *slevini,* are distinguished by a long pronotum. Within the cismontane California populations the pronotum tends to be slightly longer south of the Transverse Ranges, but some northern populations also show this tendency, particularly in the males. This general pattern of variation, with Arizona, transmontane California, and Cedros Island populations showing similar character means and marked differentiation from cismontane California populations reappears in several other characters and will be discussed below. In the Californica species group (fig. 13) the variation presents a more confusing aspect. In most cases both sexes appear at about the same point in the range of variation, but there is no obvious geographic consistency in the overall pattern.

Median pronotal width.—The cismontane California portion of the Magna group clearly breaks into a series of southern California populations with the pronotum relatively wide and a series of northern populations with the pronotum narrow (fig. 14). There is no abrupt change in this character, intermediate populations occupying the Transverse Ranges, southern Coast Ranges, and southern Sierra Nevada. This north-south cline represents a second general pattern which recurs

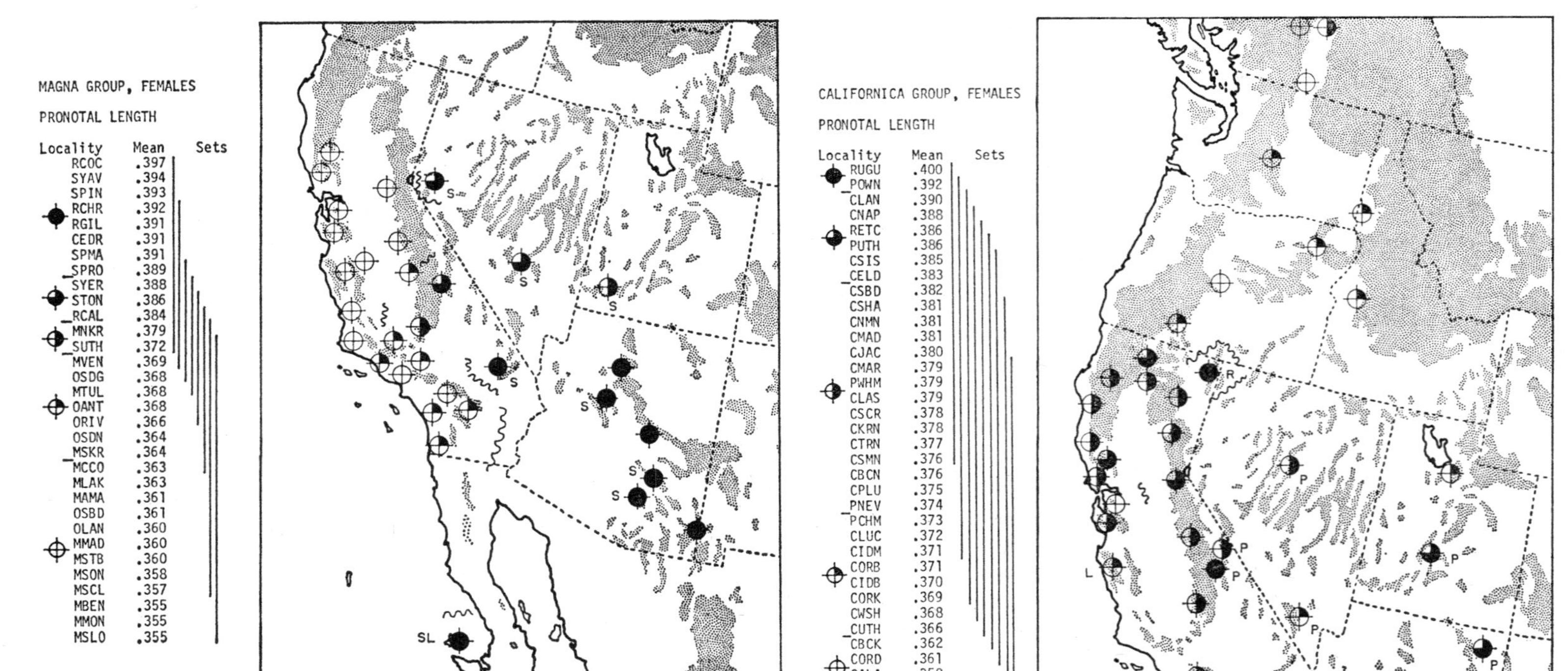

Fig. 12. Geographic variation in relative median pronotal length in the Magna species group. For explanation see text.

Fig. 13. Geographic variation in relative median pronotal length in females of the Californica species group.

in other characters of the Magna group. The remaining populations are intermediate in this character or resemble those from northern California.

Two rather sharply defined sets of populations are evident in both sexes in the Californica group (fig. 15). The Great Basin populations, with the exception of the Owens Valley, all have a narrow pronotum, and constitute the species *punctata*. Pronotal width varies considerably in the remainder of the localities, tending to be widest in the north-central Coast Ranges of California, but no abrupt changes are present. The discontinuity which occurs along the crest of the Sierra Nevada chain appears in several other characters and is discussed in the section on biogeography and evolution.

Basal pronotal width.—This character varies in the Magna group (fig. 16) in a manner similar to the last, but the montane Arizona populations are rather strongly divergent from the cismontane California components, as indicated by the small subsets near the top of the means, followed by a series of much larger subsets that contain most of the values. In regard to basal pronotal width, the three transmontane California populations are more similar to southern California *magna* than to the Arizona components of the species group; the Cedros Island sample is intermediate. The large subsets, including the lower part of the range, contain nearly all the cismontane California *magna* and most of the *sulcata* means, but the distribution maps reveal better the structure of the variation in these populations. In California *magna* again shows marked north-south polarity, with a broad range of overlap in the Coast Ranges and Sierra Nevada. The situation is less clear-cut in *sulcata*. Females of this species are distinctly larger in this character in montane Arizona, but the Yerrington, Nevada, mean is also very high, with intermediate values strung between these. In *sulcata* males (not illustrated) the highest values are also at opposite ends of the geographic range.

The Californica group means (fig. 17) fall into a few small subsets at the high end of the range, followed by much larger subsets containing most of the means. The populations which are largest in regard to this character are without exception those that are smallest in absolute body size; this includes *rugulosa*, from the Modoc plateau, *lucia* from the Santa Lucia Mountains, and the southern California samples of *californica*. These populations of small individuals resemble one another in several characters relating to stoutness of the body, but this is probably best explained as convergence, as discussed in the section on biogeography and evolution. The remainder of the Californica group shows no simple pattern of variation; although the Great Basin means tend to be lower, low values also appear in the Pacific Northwest and in central California.

Elytral length.—This is the only linearly measured character which was not converted to a ratio. In the Magna group (fig. 18) a large number of subsets each contains about one half the means, suggesting a cline. The distribution maps reveal a general pattern of this nature, with the largest values from populations in northern California, intermediates in southern California, and very small values from Arizona and Great Basin populations. There are some notable exceptions to this trend, however, such as the high means of the Riverside County and northern Baja California samples of *magna*. The Great Basin populations of *sulcata* tend

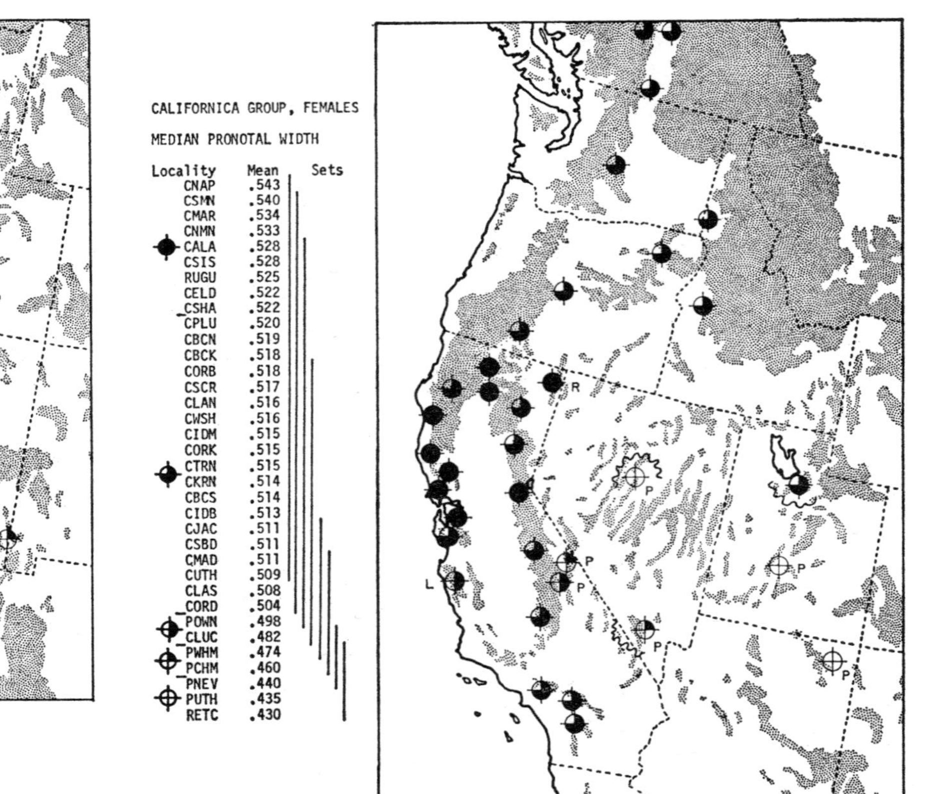

Fig. 14. Geographic variation in relative median pronotal
width in females of the Magna species group.

Fig. 15. Geographic variation in relative median pronotal
width in females of the Californica species group.

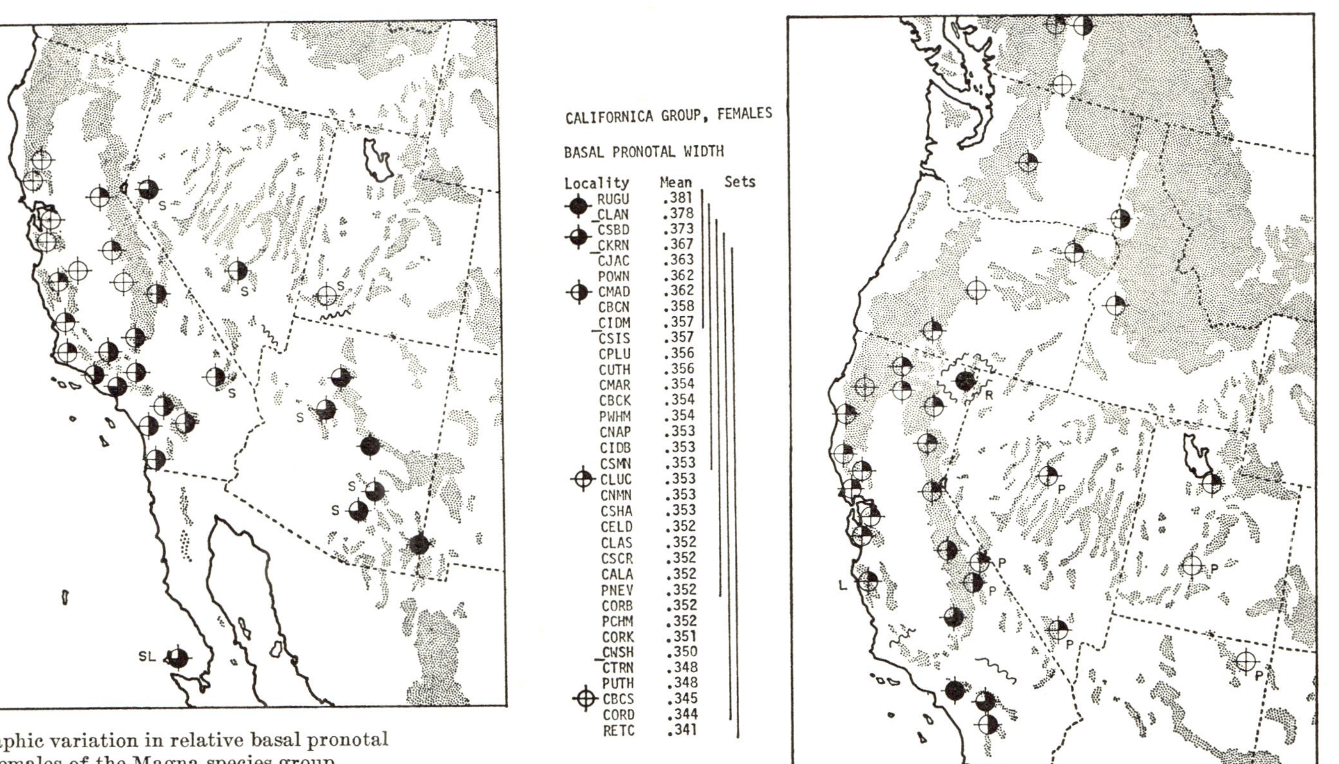

Fig. 16. Geographic variation in relative basal pronotal width in females of the Magna species group.

Fig. 17. Geographic variation in relative basal pronotal width in females of the Californica species group.

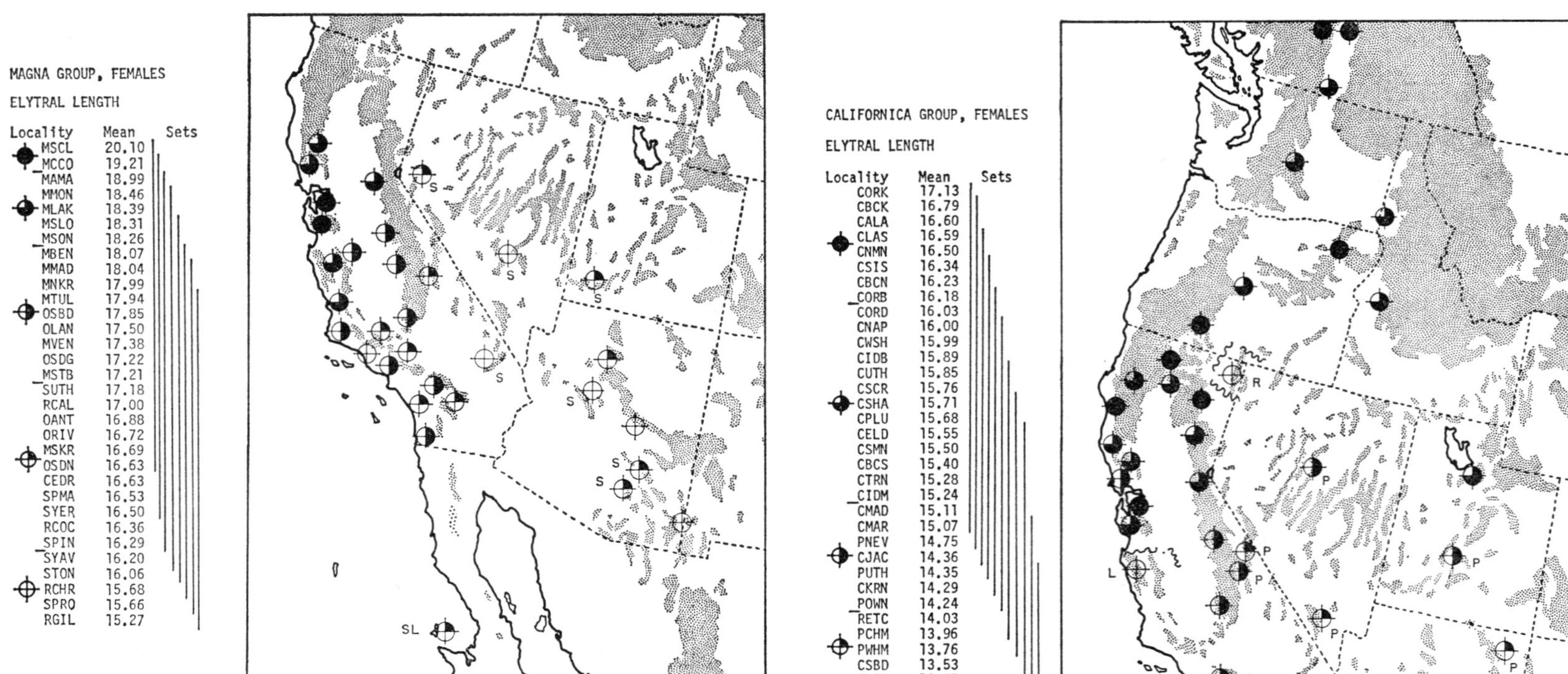

Locality	Mean	Sets
MSCL	20.10	
MCCO	19.21	
MAMA	18.99	
MMON	18.46	
MLAK	18.39	
MSLO	18.31	
MSON	18.26	
MBEN	18.07	
MMAD	18.04	
MNKR	17.99	
MTUL	17.94	
OSBD	17.85	
OLAN	17.50	
MVEN	17.38	
OSDG	17.22	
MSTB	17.21	
SUTH	17.18	
RCAL	17.00	
OANT	16.88	
ORIV	16.72	
MSKR	16.69	
OSDN	16.63	
CEDR	16.63	
SPMA	16.53	
SYER	16.50	
RCOC	16.36	
SPIN	16.29	
SYAV	16.20	
STON	16.06	
RCHR	15.68	
SPRO	15.66	
RGIL	15.27	

Fig. 18. Geographic variation in elytral length in females
of the Magna species group.

Locality	Mean	Sets
CORK	17.13	
CBCK	16.79	
CALA	16.60	
CLAS	16.59	
CNMN	16.50	
CSIS	16.34	
CBCN	16.23	
CORB	16.18	
CORD	16.03	
CNAP	16.00	
CWSH	15.99	
CIDB	15.89	
CUTH	15.85	
CSCR	15.76	
CSHA	15.71	
CPLU	15.68	
CELD	15.55	
CSMN	15.50	
CBCS	15.40	
CTRN	15.28	
CIDM	15.24	
CMAD	15.11	
CMAR	15.07	
PNEV	14.75	
CJAC	14.36	
PUTH	14.35	
CKRN	14.29	
POWN	14.24	
RETC	14.03	
PCHM	13.96	
PWHM	13.76	
CSBD	13.53	
CLAN	13.37	
RUGU	12.55	
CLUC	12.14	

Fig. 19. Geographic variation in elytral length in females
of the Californica species group.

to be somewhat larger than those from Arizona, but the differences are not significant.

Mean elytral length in the Californica group varies over a range of nearly 5 mm in both sexes and over a range of nearly 4 mm in the species *californica* (fig. 19). In contrast to the Magna group, the subsets of means are large in the upper part of the range, including the majority of samples; in the lower part of the range they become gradually smaller, containing only about one-fourth of the samples, suggesting discontinuities in population structure. The distribution maps emphasize these discordances. The very small *C. lucia*, endemic to the Santa Lucia Mountains, probably occurs sympatrically with the large-bodied *C. californica* populations in northern Monterey County. The Modoc Plateau species, *C. rugulosa*, is known to occur sympatrically with *C. californica* in the Warner Mountains. A less pronounced size difference separates the southern California montane populations from the remainder of *C. californica*, but a gradual north-south decrease in size along the Sierra Nevada is evident in both sexes. The Great Basin species, *C. punctata*, averages smaller in size than *C. californica*, especially in the western portion of its range, but the Owens Valley population is intermediate in this character, suggesting limited gene flow between *C. californica* and *C. punctata* in this area.

Elytral width.—In the Magna group this character varies in a roughly clinal manner (fig. 20), similar to *elytral length* and *median pronotal width*. The largest means are from southern California and Arizona samples and those from Cedros Island and the Little San Bernardino Mountains. The populations referred to the species *C. sulcata* repeat the pattern of slender individuals in the north and stout ones in the south, but there is no cline. The Providence Mountain *sulcata* clearly resemble the Great Basin populations rather than those in Arizona in regard to this character.

Variation in relative elytral width is much more skewed in females (fig. 21) than in males of the Californica species group, but in both sexes there is a tendency for the sets of means in the lower portion of the range to be small, while the intermediate sets are large. As with several other characters, *C. rugulosa, C. lucia,* and southern California populations of *C. californica* appear at one end of the range of variation, reflecting the short, stout bodies of these beetles, but the differences are not significant. The very low means of *C. punctata* are quite distinct, forming sets which include only a single sample of *C. californica* in either sex, but intermediacy occurs in the samples on the western periphery of the Great Basin, as in several other characters.

Abdominal thickness.—In the Magna group the pattern of variation is very similar to that of median pronotal width (fig. 14). In both sexes the means are distributed among a large number of sets, each containing about one-half the values. The transmontane California samples are similar to those from Arizona, rather than to the geographically closer cismontane California populations. The samples of *C. sulcata* break into a more slender group from the Great Basin and a stouter montane Arizona form.

In the Californica species group (fig. 22) relative abdominal thickness varies in somewhat the same manner as elytral width, but the sets of means are much

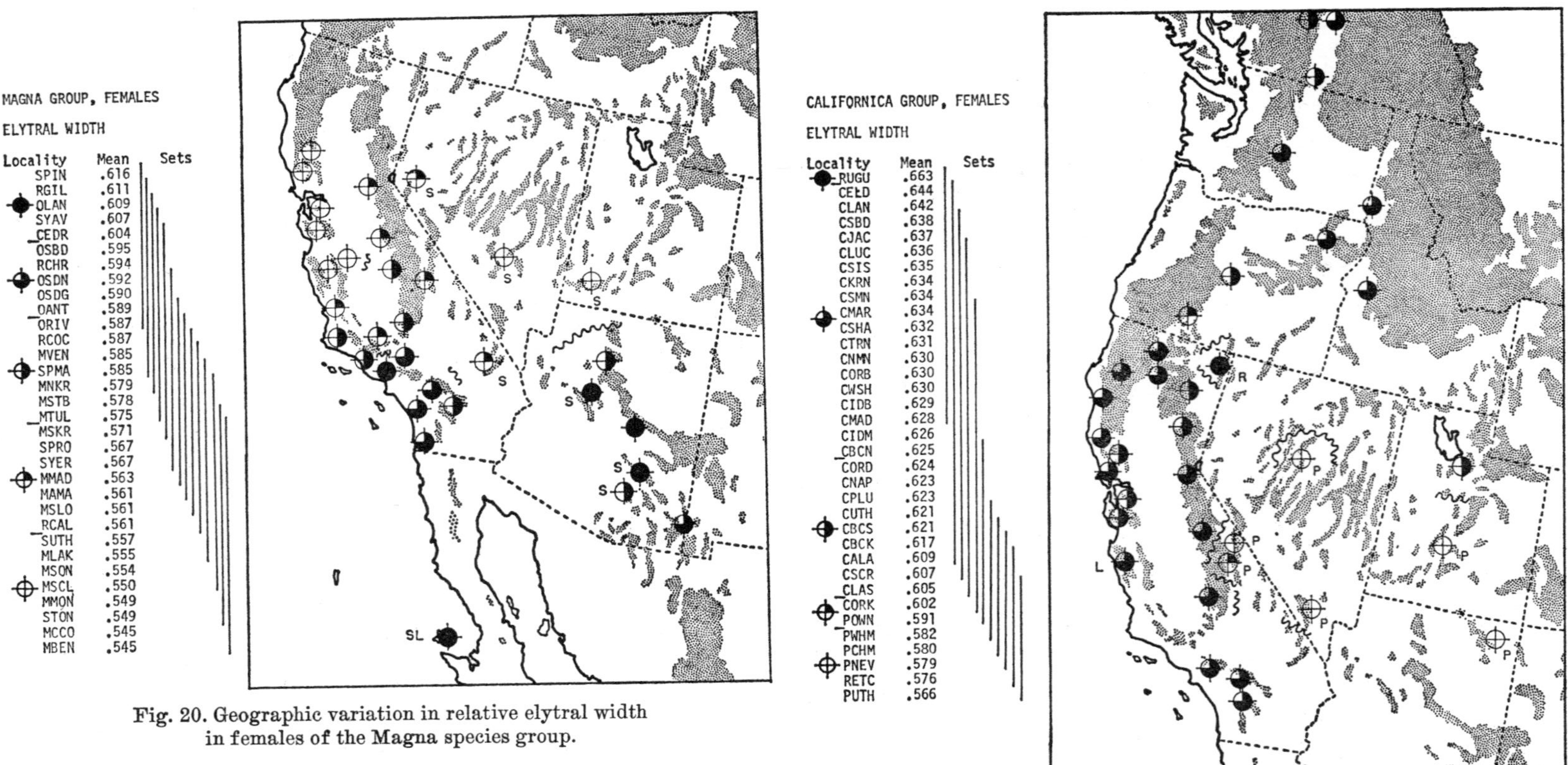

Fig. 20. Geographic variation in relative elytral width
in females of the Magna species group.

Fig. 21. Geographic variation in relative elytral width
in females of the Californica species group.

larger with the exception of the single small set at the high end of the range of morphological variation. This set contains only the very stout beetles of the Modoc Plateau, Santa Lucia Mountains, and southern California in the females, but is less restricted in the males. The populations included in *C. punctata* are relatively slender in this dimension, as in elytral width, but the means overlap broadly

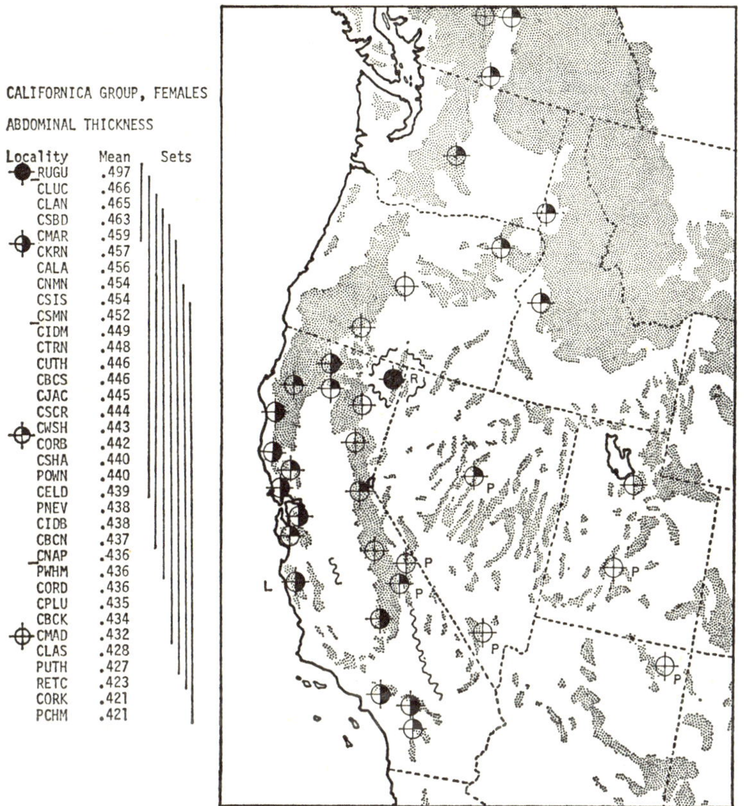

Fig. 22. Geographic variation in relative abdominal thickness
in females of the Californica species group.

with those from cismontane California and the Pacific Northwest. The variation within this last large group of samples is confusing, with no apparent major geographic trends, but it is noteworthy that both sexes of most samples have approximately the same relative value for this character.

Profemur length.—The Magna group means show a variational pattern similar to that of median pronotal width (fig. 14), with the highest values distributed through southern California and Arizona, but with gradual decrease to the north in the Great Basin as well as in cismontane California. It should be noted that with respect to abdominal thickness and elytral width the Great Basin populations of *C. sulcata* are disjunctly distinct from those in Arizona.

Profemur length varies relatively little in the Californica species group, yielding only three sets, each containing nearly all the means; the geographic distribution shows no logical arrangement of the sample values.

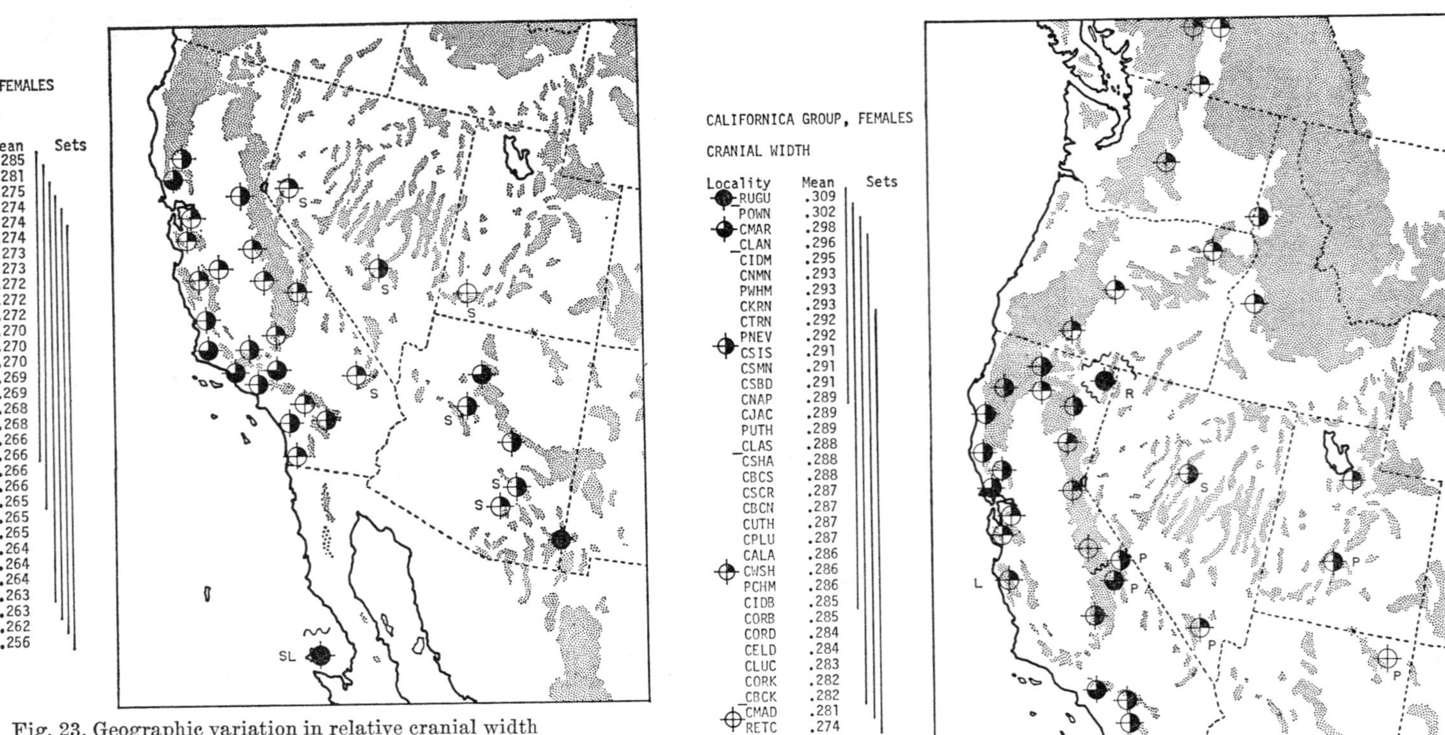

Fig. 23. Geographic variation in relative cranial width
in females of the Magna species group.

Fig. 24. Geographic variation in relative cranial width
in females of the Californica species group.

Cranial width.—The results for the Magna species group (fig. 23) resemble the variational pattern of abdominal thickness and median pronotal width, with the preponderance of high values in southern California and those of the other extreme in northern California or the Great Basin. However, the clines and disjunctions are not nearly so clear for head width; the two largest means in both sexes are those of the Cedros Island and southeastern Arizona samples, and the

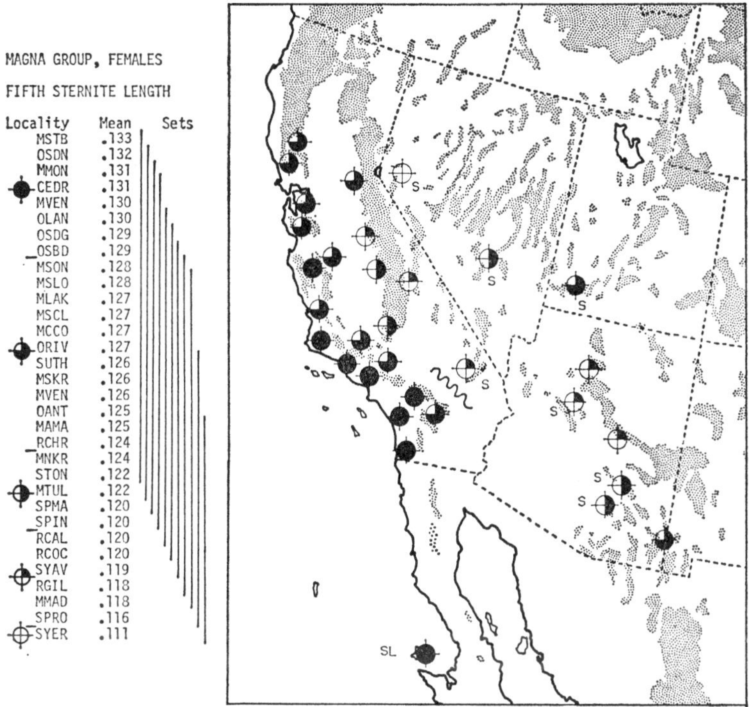

Fig. 25. Geographic variation in relative length of fifth abdominal sternite in females of the Magna species group.

females from northern California also show rather high values. The transmontane California samples show clear relationship to neither the cismontane California or Arizona populations for this character. Variation in *C. sulcata* is confused, and shows no apparent difference between the Great Basin and Arizona segments.

Relative head widths in the Californica group are evenly distrbuted over their range in males, three large sets each containing nearly all the means, with significant differences between only the populations at either extreme of variation. The distribution of sets is less uniform in the females (fig. 24), but neverthless only a few populations at the extremes are significantly different from one another. The variation revealed by the distribution maps is ambiguous but the means are rather uniformly low in the Pacific Northwest, and high in California, with many exceptions. It is noteworthy that while the head is relatively broad in the Modoc Plateau and southern California populations it is very narrow in the Santa Lucia Mountain beetles, which are otherwise stout bodied.

Length of fifth sternite.—In the Magna group (fig. 25) sternite length tends

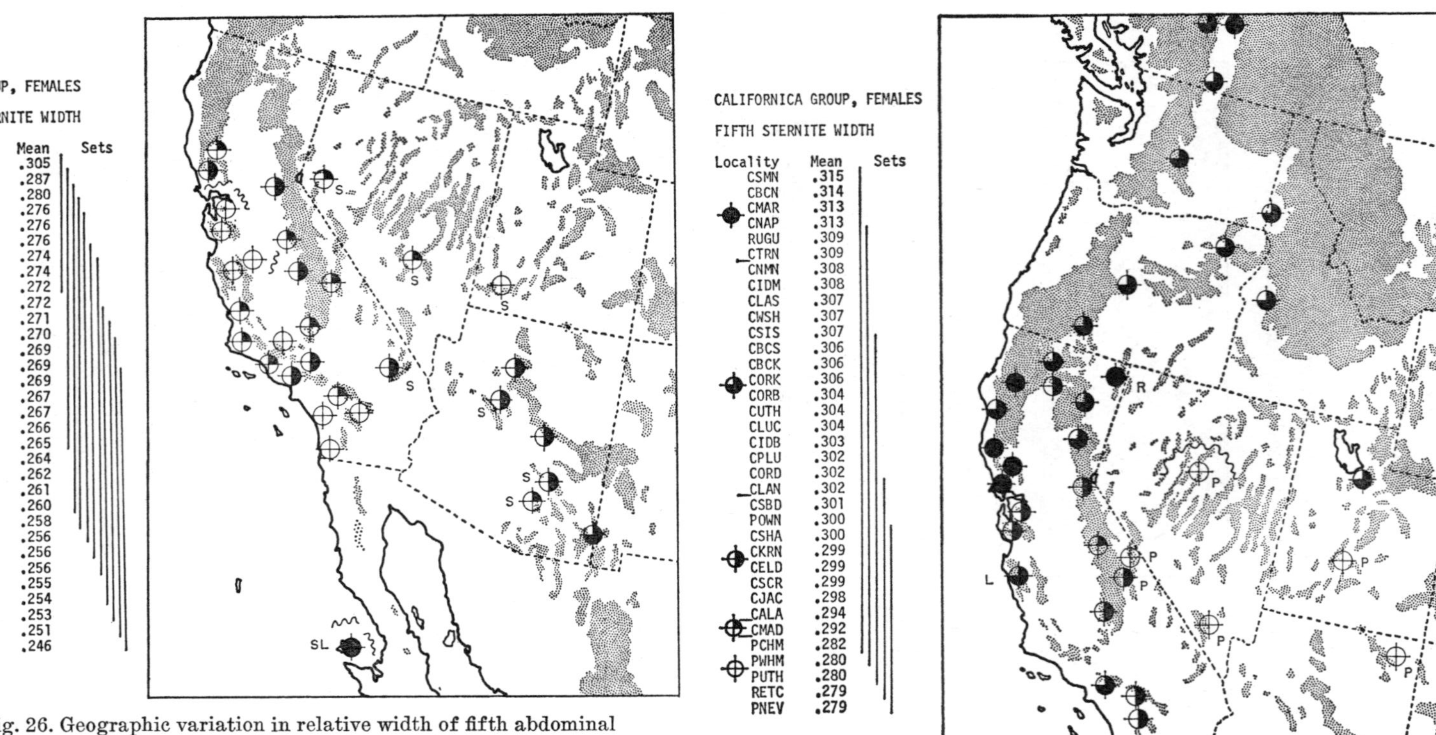

Fig. 26. Geographic variation in relative width of fifth abdominal
sternite in females of the Magna species group.

Fig. 27. Geographic variation in relative width of fifth abdominal
sternite in females of the Californica species group.

to be large in the cismontane California samples, small in the transmontane California, Great Basin, and Arizona samples, but with several exceptions, notably in the central Sierra Nevada and Cedros Island populations. The latter is similar to the cismontane California populations with respect to this character, rather than to the transmontane populations of the desert mountain ranges and those of Arizona.

Relative sternite lengths in the Californica species group (not illustrated) are all contained in a few large sets and show little significant geographical variation, although the average size tends to be smaller in the Great Basin samples (*C. punctata*).

Width of fifth sternite.—Sternite widths of the Arizona populations of *C. magna* and the Cedros Island sample fall into a small set at the high end of the morphological range (fig. 26); the remainder of the means fall into larger classes with few significant differences. In the species *sulcata* a tendency toward a narrower sternite in the Great Basin is evident but not pronounced.

In the Californica group (fig. 27) most of the Great Basin, southern California montane, and southern Sierra Nevada means are contained in a small set of low values, with the remaining samples included in a series of larger sets. It will be noted in regard to this character as well as several others that the Great Basin populations resemble those from southern California, but this does not imply an evolutionary relationship. The southern California populations unquestionably are closest to those from the southern Sierra Nevada, while *C. punctata* forms a distinct isolate in the mountains of the Great Basin, with a narrow band of intermediates along the eastern escarpment of the Sierra Nevada.

Elytral rugosity.—The means of the Magna group (fig. 28) fall into a small set of high values and several larger sets of low values. The populations of *sulcata* are mostly more rugose than those of *magna* and *slevini* (Cedros Island), but there are several exceptions. In cismontane California the most rugose samples are from the north. There are no consistent geographic trends.

Elytral rugosity is highly variable in the Californica species group (fig. 29), characterizing the populations of several geographic regions. The Great Basin populations (*punctata*) are uniformly of low rugosity, with the Owens Valley sample again intermediate between *punctata* and the more rugose *californica*. Within the large group of populations in California and the Pacific Northwest, those of the San Francisco Bay region stand out because of their extremely low rugosity; they are very similar to the California populations of *C. magna* in this character, as well as in several punctation characters, as shown below. To the north these populations are clinally connected to the more rugose beetles of the Siskiyou Mountains, the cline continuing irregularly around the central valley of California and south along the Sierra Nevada, terminating with the very rugose populations of the southern California mountains. The samples from extreme northern California and the Pacific Northwest are nearly uniform in this character, most of the means falling into the large sets of intermediate values. The Modoc Plateau sample (*rugulosa*) is an exception, with the highest values recorded for both sexes. The Santa Lucia Mountains population differs from the others of small body size in being only slightly rugose, resembling *punctata* in this character.

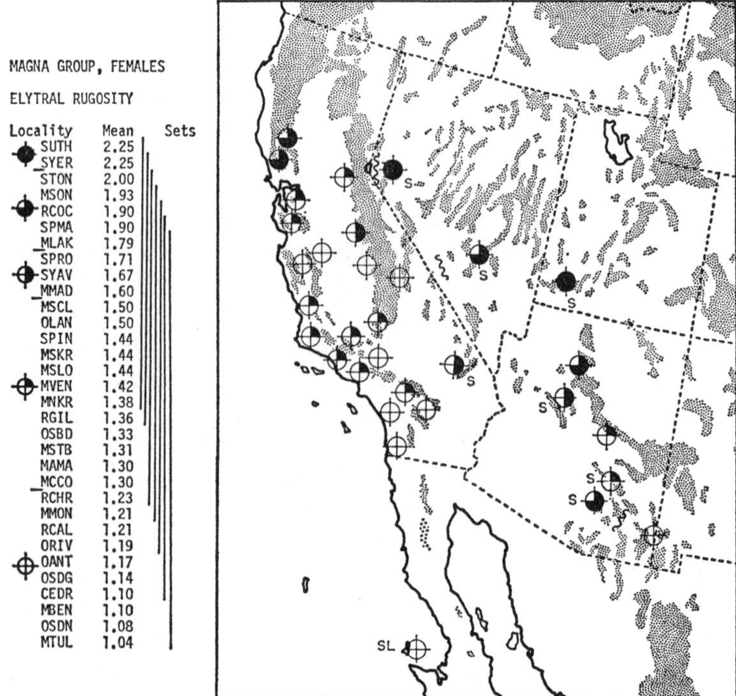

Fig. 28. Geographic variation in elytral rugosity in females of the Magna species group.

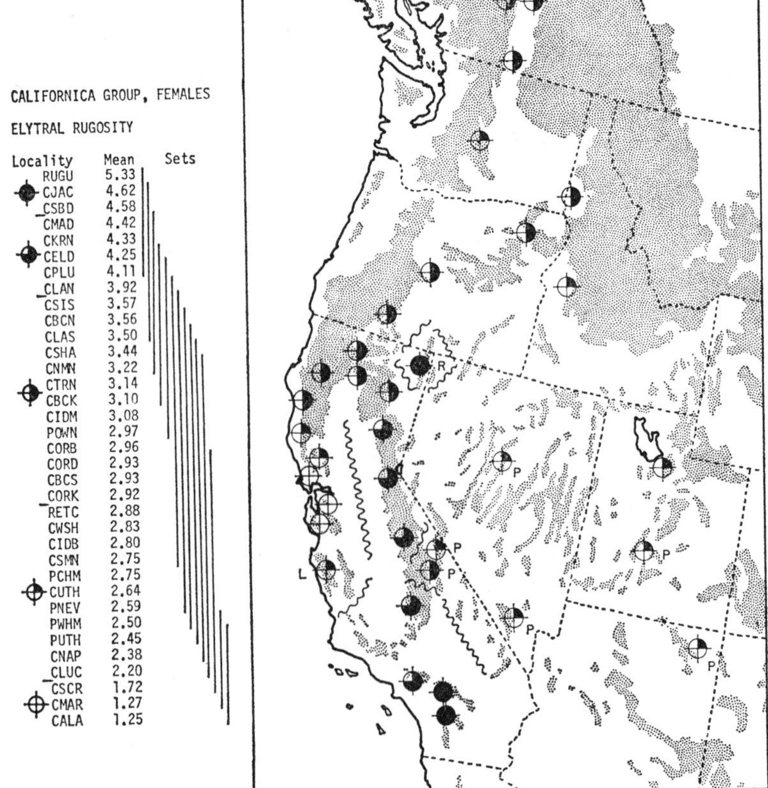

Fig. 29. Geographic variation in elytral rugosity in females of the Californica species group.

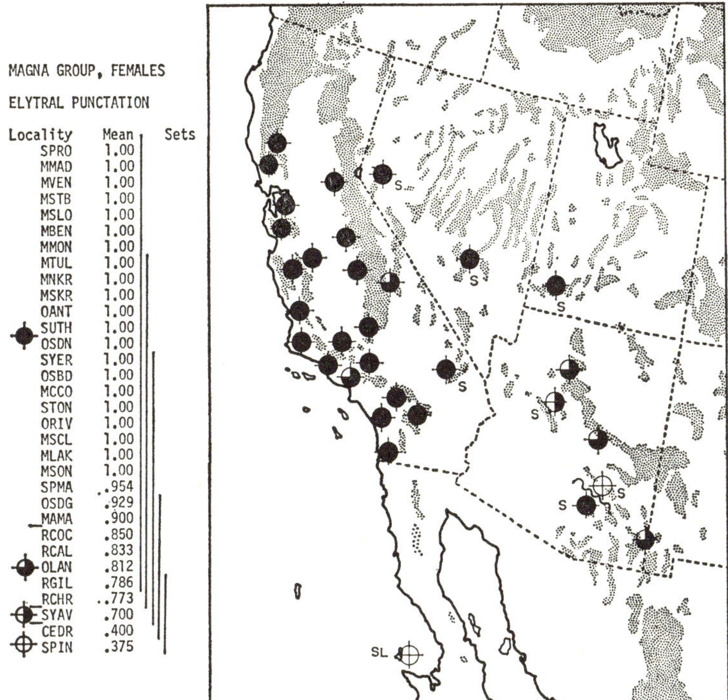

Fig. 30. Geographic variation in elytral punctation in females of the Magna species group.

The four punctation characters show very similar variation, reflecting the finely, sparsely punctate surfaces of beetles in the Magna group and the more coarsely, densely punctate cuticle of those in the Californica group. For this reason these four characters are discussed together for each species group. The STP diagram and distribution map for elytral punctation in the Magna group are shown in figure 30. The punctation characters are of chief interest in this species group because of the distinct tendency of the Cedros Island sample (*C. slevini*) to appear at the low extreme of variation. The punctation tends to be fine in the Arizona samples, but is variable. The Ventura County sample of *magna* has an unaccountably low value, probably due to a sampling error.[3] The values for the Arizona populations of *C. sulcata* are frequently lower than for those of the Great Basin and Providence Mountain populations, but the differences are usually insignificant. The distribution of the means of the cismontane California samples show no consistent geographic patterns and practically no significant differences.

In the Californica species group punctation is highly variable, showing few obvious geographic patterns; punctation of the hypomeron shows no significantly different subsets and is not discussed further. Each of the other three characters (punctation of elytra, head, and pronotum) show a moderate number of homogeneous subsets with significant differences mostly between pairs of means at opposite extremes of the variation. With regard to elytral punctation (fig. 31)

---

[3] Specimens collected subsequent to this analysis show that punctation is variable throughout the transverse ranges.

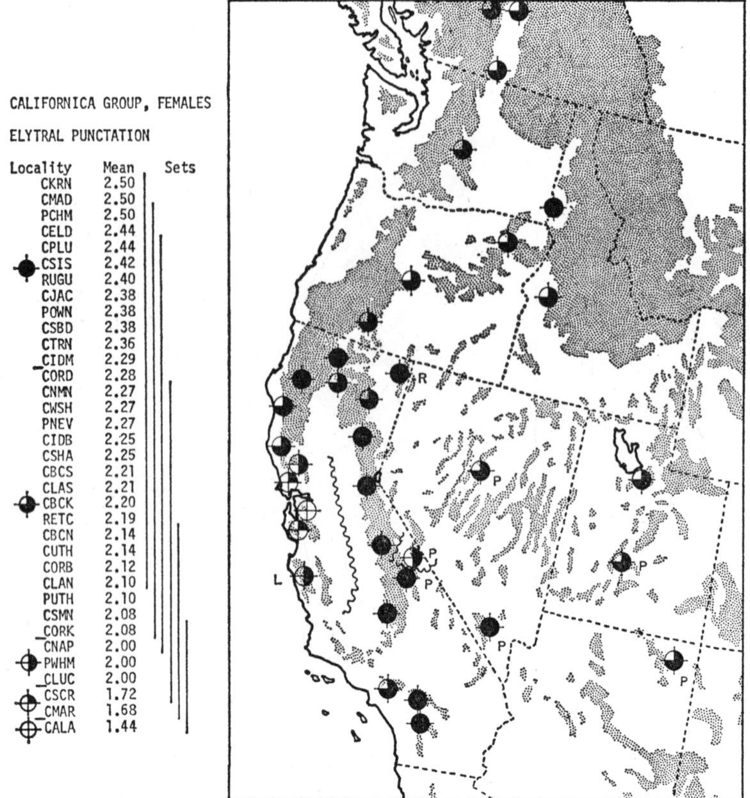

Fig. 31. Geographic variation in elytral punctation in females of the Californica species group.

the San Francisco Bay region populations stand out because of their low means, clinally connected with the populations to the north; this pattern holds, although less consistently, for pronotal and cranial punctation as well (fig. 32). Punctation is consistently very coarse in *C. rugulosa* (Modoc Plateau), and tends to be coarse in the Sierra Nevada and British Columbia samples, but there are numerous exceptions, particularly in the former region. The populations of *C. punctata* (Great Basin) are highly variable in all the punctation characters, belying the name of this species.

Prosternal rugosity.—There are no significant differences in the Californica species group. In the Magna group (fig. 33) the variation is similar to that of gular rugosity and the punctation characters, but the means are more evenly distributed, with a smaller gap between *C. slevini* and the least rugose samples of *C. magna*. In both sexes the Great Basin samples of *sulcata* have higher means than the Arizona samples and in cismontane California the most rugose populations of *magna* are concentrated near the coast. There is also a tendency for the central coastal populations of California *magna* to be weakly rugose with highest rugosity in the San Francisco Bay region and Transverse Ranges.

Gular rugosity.—This character shows significant differences only in the Magna

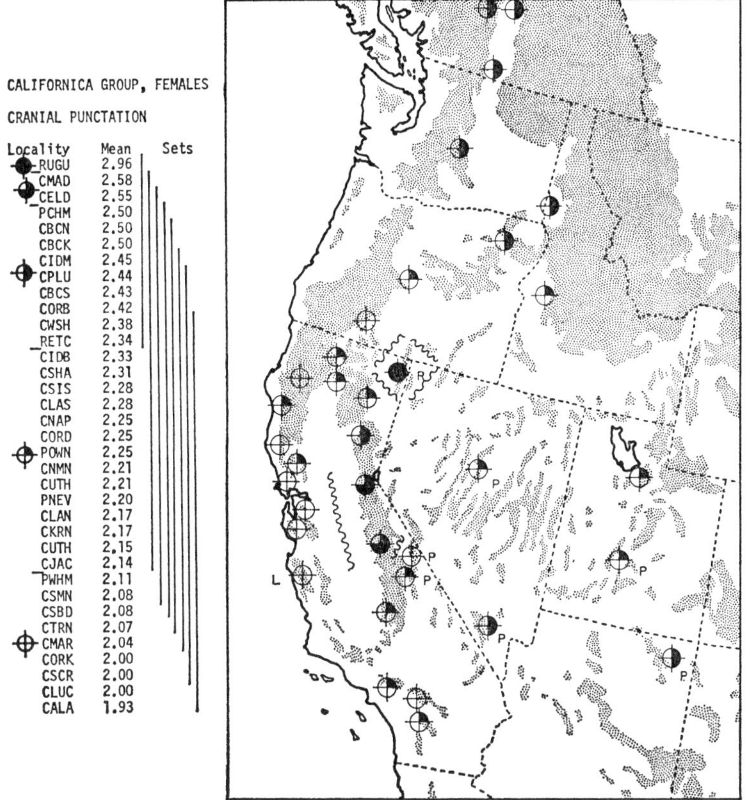

Fig. 32. Geographic variation in cranial punctation in females
of the Californica species group.

species group. The pattern of variation is similar to that of the punctation characters (fig. 30), particularly in the very low values for both sexes of *C. slevini* (Cedros Island), but the Arizona populations are not detectably different from those in California. In the males, but not in the females, a definite coastal distribution of high values is evident in *C. magna,* and the means of the Great Basin samples of *sulcata* are slightly higher than those of Arizona *sulcata,* but these differences are not significant.

Prosternal process.—In the Magna species group (fig. 34) there is little significant variation in this character except at the lower extreme of variation. In both sexes the prosternal process is declivous (low means) in the Cedros Island population. It is prominent (mean values greater than 2.50) in nearly all the remaining samples, with exceptions in central Nevada (possibly a sampling error) and northern California.The largest values are concentrated in southern California (both sexes) and Arizona.

The distribution of homogeneous sets is sharply skewed toward the upper end of the range of variation in the Californica species group (fig. 35) and the differences in expression of this character show distinct geographic patterning. The highest means (prominent prosternal process) are those of the southern California

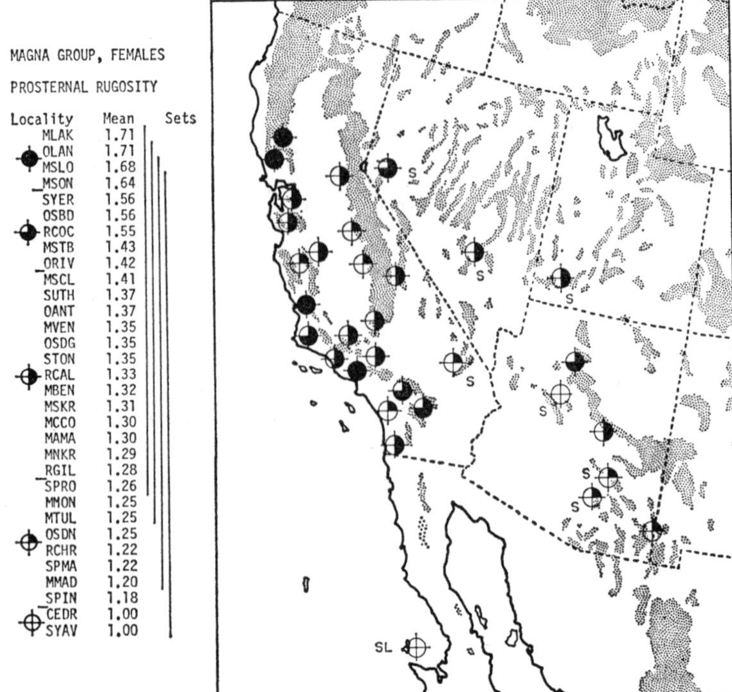

Fig. 33. Geographic variation in prosternal rugosity in females of the Magna species group.

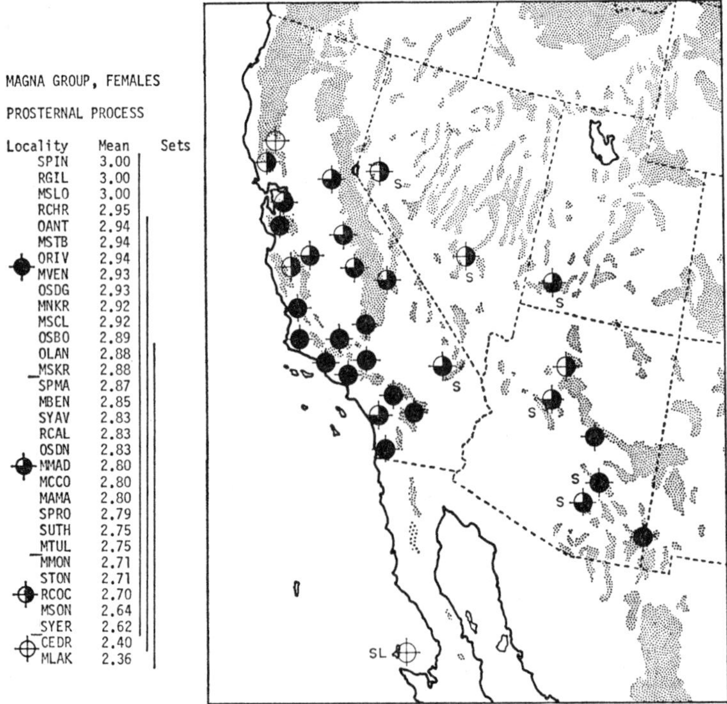

Fig. 34. Geographic variation in shape of the prosternal
process in females of the Magna species group.

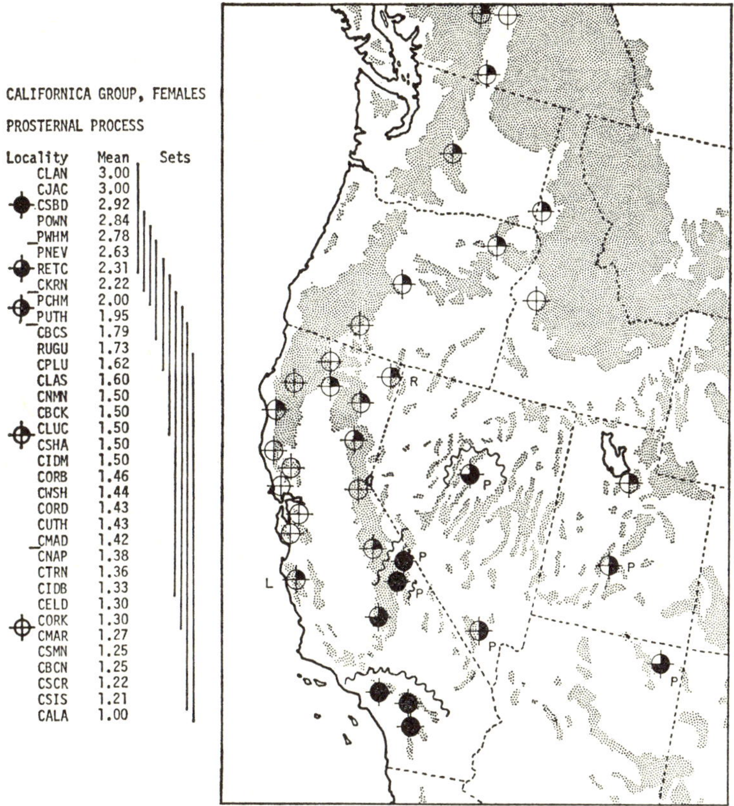

Fig. 35. Geographic variation in shape of the prosternal
process in females of the Californica species group.

montane samples, with gradual decrease to the north along the Sierra Nevada;
mean values are also large in the Great Basin (*C. punctata*), especially in the
White Mountains and Owens Valley. The Modoc Plateau and Santa Lucia Mountain populations (*C. rugulosa* and *C. lucia*) are near the upper extreme in variation but are not significantly different from most of the populations of *californica*,
which have the prosternal process uniformly declivous.

Curvature of dorsum.—The elytral dorsum is slightly more convex in males
than females in both species groups. In the Magna group only two or three large
sets each contain nearly all the means, with significant differences only between
the most extreme values. In both sexes the dorsum tends to be most convex in
cismontane southern California and Arizona, and in the males on Cedros Island.

In the Californica species group (fig. 36) one or more small sets of high means
is followed by several large sets containing most of the samples. The distribution
map reveals a pattern similar to that for elytral rugosity, with the San Francisco
Bay region marking one end of an irregular cline around the northern end of
the central valley and south along the Sierra Nevada. In the central Sierra, where
a pronounced flattening of the elytra occurs, the cline is disrupted; in the Greenhorn Mountains at the southern tip of the Sierra and in the Transverse Ranges,

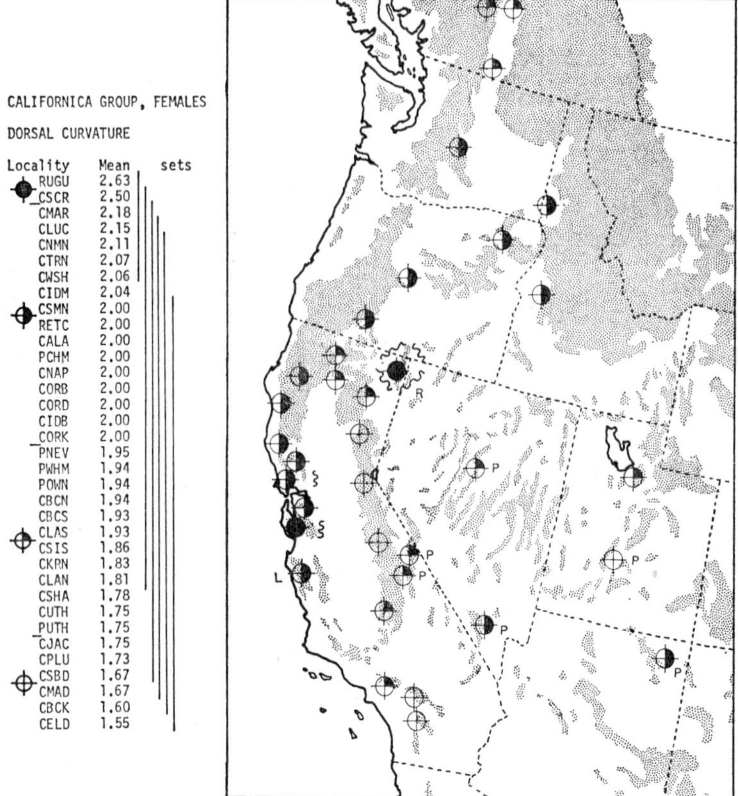

Fig. 36. Geographic variation in curvature of elytral dorsum
in females of the Californica species group.

flattening is not so evident. In the Pacific Northwest and Great Basin the beetles tend to be rather uniformly, weakly convex in the females, moderately and variably so in the males.

Elytral sulcation and color.—These characters are practically invariant throughout most of the genus. In the Californica species group only the Colorado Plateau populations evidence any tendency toward elytral sulcation; the elytral sculpturing in most individuals is both sulcate and coarsely, transversely rugose, producing a cancellate appearance. The same populations are also unique in the reddish-black elytra of many specimens. For both these characters, analysis of variance segregates the Colorado Plateau specimens into a single member set which is mutually exclusive with a single large set containing the remaining samples; this is the only occurrence of completely nonoverlapping sets in the study. However, the scarcity of collected material from the Colorado Plateau necessitated combining specimens from a large geographic area into a single sample, concealing a gradual change between the exceptional populations of the plateau and the typical *punctata* of the Great Basin. The few available specimens from south-central Utah are distinctly intermediate in appearance. Several other species of tenebrionids display a marked convergence toward reddish color and can-

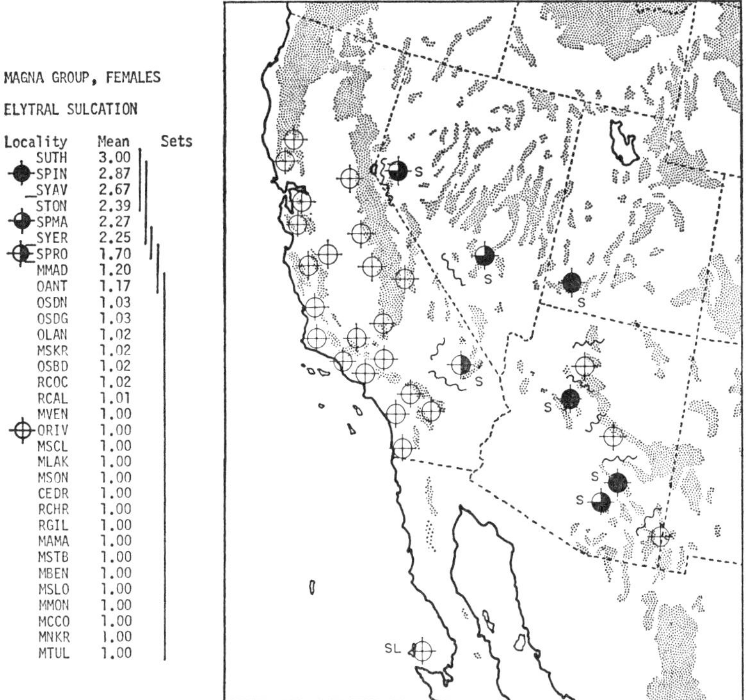

Fig. 37. Geographic variation in elytral sulcation in females of the Magna species group.

cellate elytra in the Colorado Plateau segment of their range, a point which will be considered in more detail later.

In the Magna species group the highest means for elytral sulcation (fig. 37) are consistently from Great Basin and montane Arizona samples, with a slightly sulcate population in the Providence Mountains, California. Collectively these populations are referred to by the name *C. sulcata,* although the Arizona populations are more similar to Arizona representatives of *C. magna* in most other characters, a factor which will be elaborated in the biogeography section.

Color variation in the Magna group is confined to the Arizona populations of *C. sulcata* and the sample of *C. magna* from Coconino County. The high mean of the Sonoma County sample is probably explained by the presence of one or more teneral individuals, and most of the Arizona samples are not significantly different from the bulk of the samples contained in the single large set of low values. The Yavapai County population, on the other hand, is significantly different from almost all others in the study, most closely resembling the Colorado Plateau populations of the Californica species group in color.

Summary of Univariate Analysis

Analysis of morphological variation reveals patterns of gradual change in most characters, sometimes as simple clines. Abrupt discontinuities are infrequent. In each species group certain constellations of characters present closely congruent geographic trends, apparently reflecting general adaptations of body proportions or cuticular sculpturing. The north-south clinal transformation from a slender to a relatively stout body form in cismontane California populations of *C. magna* is a notable example, involving elytral length and width, fifth-sternite length and width, and median and basal pronotal width. Although the ecological rules frequently invoked to explain morphological variation in vetebrates have been applied to insects and other invertebrates (Ray, 1960), the present discussion is limited largely to a comparison and analysis of the geographic patterns of the various characters. The ecological factors which may underlie these patterns are mentioned where obvious correlations seem to exist. The general geographic patterns of variation examined here will be discussed in terms of taxonomy, historical geology, and paleogeography in appropriate sections below.

The Magna group.—This series of populations of predominantly southern distribution is characterized by smooth cuticular surfaces with reduced punctation and sculpturing, moderately curved abdomen and elytra, and prominent, horizontal prosternal process, with two exceptions. First, the elytral sculpturing is moderately rugose and the punctation moderately coarse in the Santa Catalina Island individuals (known from four females; not included in univariate analysis), which are very similar to cismontane southern California specimens of *C. magna* in other characters. Second, the elytral intervals are raised into sulci in *C. sulcata,* which shows only insignificant differences from *C. magna* in other characters. The Magna species group is also distinguished by a relatively long, narrow fifth abdominal sternite, narrow cranium, and short legs, although these differences are not apparent without measurement. There is slight overlap between the Magna and Californica species groups in all these characters.

Within the Magna species group several recurrent geographic patterns of character change are evident. The most obvious is the previously described parallel clinal variation in six measurements of body proportion in cismontane California populations of *C. magna.* In these six characters the variation within this series of clinally connected populations is nearly as great as that in the entire species group. Comparison of distribution maps shows that the clines usually extend as far south as the Los Angeles Basin; Peninsular Range populations resemble those just north of the Transverse Ranges in these characters. No single population in this series differs significantly in any character from its geographic neighbors, but there frequently are significant differences between the geographic extremes.

A second series of populations from montane Arizona, including both *magna* and *sulcata,* is characterized by smaller size, the shape of the pronotum (relatively high pronotal length and basal pronotal width), and reduced elytral rugosity and punctation. In elytral width and relative femur length the Arizona specimens are most similar to the southern California form of *magna,* but in median pronotal width are similar to the northern California form. The populations isolated in the Owens Valley region and the desert mountain ranges are intermediate in size between the California and Arizona forms of *magna,* similar to the former in basal pronotal width and to the latter in pronotal length. Three males and three females from the Sierra San Pedro Martir (not included in univariate analysis) show similar relationships. The Cedros Island population is extreme in many characters, resembling the Arizona form in some and the California form in others.

The populations designated *sulcata* differ from *magna* principally in the single character of elytral sulcation. The sulcate and nonsulcate populations in Arizona share several characters (elytral width, basal pronotal width, elytral punctation) which differentiate them from the more slender bodied Great Basin populations of *sulcata*. It is of interest that in California the two body dimensions are also relatively small in the north and large in the south, suggesting a possible polyphyletic origin for *C. sulcata*. This problem is discussed in more detail later. The Providence Mountain population is very weakly sulcate (along with two specimens from Inyo County, California), with similarities in other characters to both the Arizona form (pronotal length) and the Great Basin form (elytral width).

The Californica group.—This is a northern group of species, distinguished by coarsely punctate cuticular surfaces, strongly rugose postoral region, rugose elytra, posteriorly downcurved abdomen, and usually declivous, flattened prosternal process. The San Francisco Bay region populations are singularly smooth and finely punctate, strongly resembling *C. magna* in these characters, and specimens from southern California have the prosternal process prominent and horizontal, again typical of *magna*.

Within the Californica species group two major series of populations are evident. The cismontane California, Pacific Northwest, and Wasatch Mountains (Utah) beetles are characterized by relatively broad elytra and pronotum, giving them a squat appearance. These populations are extremely variable in size, elytral rugosity, and dorsal punctation, but show very few abrupt discontinuities. The most distinct is the Modoc Plateau population, which not only shows highly significant differences in elytral length and abdominal thickness but is extreme in most other characters as well. The relatively stout body dimensions of the Modoc Plateau beetles are shared by the small-bodied specimens of the Santa Lucia Mountains and those of the mountains of southern California; the latter also resemble the Modoc Plateau individuals in being very strongly rugose. These three populations are well differentiated by other characters, indicating that their similarities are explicable as a result of convergence to small body size. Reduced size and stout proportions may be an adaptation to the relatively dry areas inhabited by these beetles. However, the populations of the much drier Great Basin combine small size with very slender proportions and the populations of the relatively arid San Francisco Bay region are large bodied.

The populations of the Pacific Northwest are nearly homogeneous in all characters, but in California significant variation occurs in several features. The irregular clines extending north from the San Francisco Bay region to the Siskiyou Mountains, then south along the Sierra Nevada chain, and finally terminating in the mountains of southern California have already been mentioned. Gradual changes in elytral rugosity involve this entire segment of the distribution, but variation in dorsal punctation continuously increases only to the terminus of the Sierra Nevada, decreasing in southern California. The central Sierran populations are singular in the weakly flattened elytral dorsum; southern Sierran specimens are frequently of small size and occasionally show the prominent prosternal process typical of southern California material.

A *rassenkreise* around the central valley was first detailed in *Ensatina eschscholtzi* (Gray) (Stebbins, 1949) but this is a distributional pattern common to several other reptiles and other transition-zone organisms in California (see Stebbins, 1954, for examples). Typically, as in *Coelocnemis,* this pattern of distribution incompletely encircles the central valley, with gaps at the northern or southern end of the valley, depending on the organism. In *Coelocnemis* a small distributional gap exists between the Transverse Ranges and the Sierra Nevada and another larger disjunction occurs in the dry, relatively low ranges south of the Santa Lucia Mountains. It is not clear whether the Santa Lucia populations belong to the circum-valley cline. In size and body proportions they are similar to southern California specimens, but do not share their coarse rugosity or prominent prosternal process.

The second major series of populations of the Californica species group occurs in the Great Basin and on the Colorado Plateau. These populations, all included in the species *C. punctata,* are distinguished by relatively small size, slender body proportions (elytral width, median and basal pronotal width), and moderately prominent prosternal process. The Colorado Plateau specimens differ in their peculiar reddish-black color and cancellate elytra, but, as previously mentioned, intergrade with typical Great Basin specimens in southeastern Utah. These color and sculpturing peculiarities may be an adaptation to the red, sandy soil conditions of the Colorado Plateau. *Eleodes obscura* Say is represented by the subspecies *dispersa* LeConte in this area and similar geographic differentiation occurs in other species of *Eleodes,* and in the genera *Gonasida* and *Stenomorpha* where separate Colorado Plateau species or subspecies are recognized. The Arizona populations of *C. sulcata,* particularly those from Yavapai County, also show a distinct reddish cast to the elytra.

Along the eastern escarpment of the Sierra Nevada and in the Owens Valley the beetles are intermediate in every character separating *californica* and *punctata.* The White Mountains sample of *punctata* is also slightly biased toward the *californica* phenotype. A few other specimens from the periphery of the Great Basin (Carson River Canyon, Alpine County, California, two individuals; Lassen County, California, two individuals; Craters of the Moon National Monument, Idaho, six individuals) also tend to be intermediate in these characters, but were not included in univariate analysis. In central Utah, where *californica* and *punctata* occur in close sympatry, no intermediacy is recognizable.

PHENOGRAMS (ANALYSIS)

The relationships examined here are measures of the averages of similarities of the 21 characters considered above. For analysis of overall phenetic relationships all the OTUs of both species groups are considered together. In the phenograms shown earlier the locality samples were treated as characters; here they are the OTUs, each with 21 characters. The extra samples are those with fewer than four individuals per sex (see Appendix 1). The mean character values for each sample appear in the cells of the data matrix. Phenograms based on Pearson's product-moment correlation coefficient and the taxonomic distance coefficient (see

Sokal and Sneath, 1963) were constructed for each sex, using unweighted, variance-standardized data.

The correlation coefficient phenogram for 81 female samples is shown in figure 38. The cophenetic correlation coefficient is 0.916, indicating relatively close agreement between the similarity matrix and the phenogram. The samples split into a pair of very large branches, each with about one-half the OTUs. These major branches represent the Californica and Magna species groups and correspond very closely with the groups that are indicated by the ecological and biochemical evidence. The most poorly placed samples are CAPE (Cabo San Lucas, Baja California Sur) and MMRT (Sierra San Pedro Martir, Baja California Norte). Neither of these samples were included in univariate or biochemical analysis because of lack of material, but at least the latter clearly belongs to the Magna species group. The Cabo San Lucas population is known from only two specimens which combine morphological characters from both species groups, and its position on the phenograms is variable. However, the zoogeographic evidence presented later indicates that it belongs to the Magna species group.

The Magna species group, represented by the upper half of the phenogram, consists of two large branches representing (1) the California localities, and (2) the montane Arizona–Great Basin localities. Santa Catalina Island and southern Kern County are relatively isolated on this phenogram. It should be noted that the California branch contains all but two of the cismontane samples (the Kern County samples are exceptions) and that the transmontane California samples cluster with those of montane Arizona. The relationships of these transmontane samples were ambiguous on the basis of univariate analysis, showing similarities to both the Arizona and California populations. The Cedros Island sample, which seems distinct on the basis of many single characters, also clusters with the Arizona populations. These relationships assume greater significance in the light of the reconstructed zoogeographic history of the genus, discussed at length later. As expected, *sulcata* and the Arizona populations of *magna* appear in separate clusters, but the Yavapai and Coconino County samples form a subcluster of their own, intermediate between the *sulcata* and *magna* clusters. Apparently this is caused largely by the reddish coloration of specimens from these localities, since cluster analysis with the single color character omitted clearly separates *sulcata* and the Arizona samples of *magna*. The relatively minor differences between Arizona and Great Basin populations of *sulcata* are not clear in the phenogram.

The most deceptive linkages in the entire phenogram involve the populations related by multiple-character clines. The cismontane California populations of *magna* are consistently split into northern and southern clusters by all methods of hierarchical analysis. These clusters are not well separated at lower levels of similarity and their composition varies between sexes and methods of analysis, suggesting distortion, but it is not possible to recognize the clinal structure from the phenograms. It is worth noting that within the clusters of northern and southern *magna* samples the relationships of individual OTUs frequently coincide with geographic contiguity. This is particularly evident in the series of northern localities where distinct subclusters of north coastal (MSON, MLAK, MGLN), central

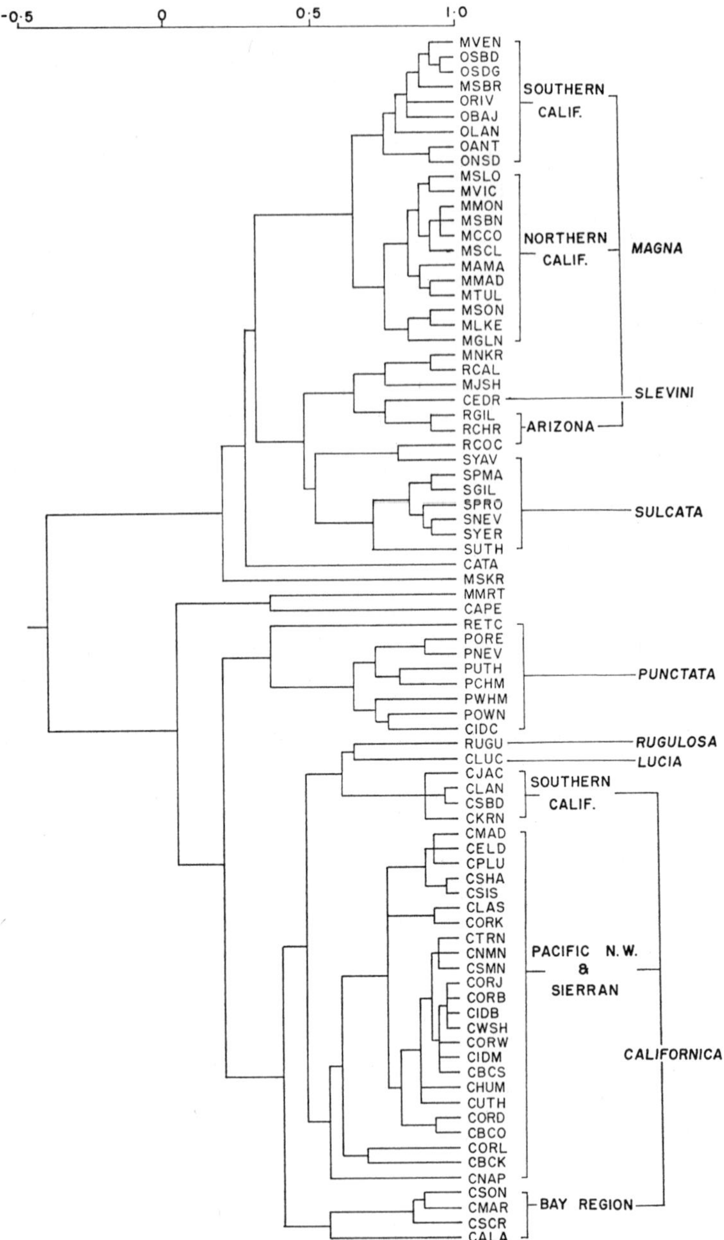

Fig. 38. Correlation phenogram of female samples. Vertical lines on the right side of the phenogram indicate clusters of samples discussed in the text. Names of species are italicized. Infraspecific groups of samples are indicated by roman type.

coastal (MMON, MSON, MCCO, MSCL), and Sierran (MAMA, MMAD, MTUL) populations are linked at high levels of similarity.

The large number of samples of the Californica species group are arranged in very complex clusters. The anomalous position of CAPE and MMRT have already been mentioned. The remainder of the samples are split into a small cluster containing the *punctata* samples and a much larger cluster including the *californica* samples as well as *rugulosa* (RUGU) and *lucia* (CLUC). The phenogram depicts *punctata* as being quite distinct from the other species, with the single Colorado Plateau population (RETC) relatively isolated. These results reinforce the relationships suggested by univariate analysis, but subsequent analysis shows that the phenetic relationships of *punctata*, particularly the Colorado Plateau population, are somewhat ambiguous. The large cluster containing nearly all the *californica* samples consists of three major subclusters, the first linking the southern California samples (including Kern County) with those from the Modoc Plateau (RUGU) and the Santa Lucia Mountains (CLUC). These are the populations with small average size and stout body proportions. A second group includes the San Francisco Bay region samples, distinguished by reduced punctation and rugosity, but clinally connected in these characters with the populations to the north according to univariate analysis. The distortion of actual phenetic relationships is much greater here than in the *magna* cline discussed above. The rest of the samples are contained in a single large cluster, but the linkage relationships again reflect geographic contiguity. Thus, a Sierran-Siskiyou Mountain group (CMAD, CELD, CPLU, CSHA, CSIS), a Modoc Plateau group (CLAS, CORK), and a north coastal California group (CTRN, CNMN, CSMN) of populations may be distinguished. It is not clear why the three samples, CORL, CBCK, and CNAP, are relatively isolated.

The taxonomic distance phenogram for females is shown in figure 39; the cophenetic correlation coefficient is 0.880. The skewness of the branching relationships toward the base of the phenogram is typical of phenograms based on taxonomic distance (Rohlf and Fisher, 1968). While the broad outlines of the phenograms based on correlation coefficient and taxonomic distance are very similar, there are many discordant details. An obvious difference is that several of the OTUs which were linked to clusters at a low level in the correlation phenogram are even more isolated according to taxonomic distance. This is most evident with regard to MSKR, RUGU, and SYAV. In contrast, CAPE, CATA, and MMRT, which were isolated in the correlation phenogram, cluster with the montane Arizona samples in the distance phenogram. It may also be noted that the Colorado Plateau sample (RETC) of *punctata*, the Alameda County sample (CALA) of *californica*, and the Santa Lucia Mountains sample (CLUC) are removed from their respective correlation clusters to more isolated positions in the distance phenogram.

In the Magna group the cluster of southern OTUs has been enlarged by the addition of MSLO and three samples (MNKR, MNYO, and MJSH) which previously clustered with the montane Arizona localities. These changes reflect the deceptive portrayal of the north-south cline in the former instance and the

Fig. 39. Taxonomic distance phenogram of female samples.
See figure 38 for further explanation.

phenetically intermediate position of MNYO and MJSH in the latter. The distance phenogram emphasizes the phenetic integrity of *sulcata,* with the exception of the Yavapai County sample.

The taxonomic distance branching relationships in the Californica species group are more complex than in the correlation phenogram, but the same clusters are discernible. The tendency of populations from geographically contiguous localities to cluster together is again evident. The differences that occur are minor, such as the inclusion of the Sierran samples (CMAD, CELD) in a cluster with one of the British Columbia samples (CBCK).

Correlation and distance phenograms of the 81 male samples are shown in figures 40 and 41; the cophenetic correlation coefficients are 0.925 and 0.878, respectively. The results for males match those for females quite closely and only the obvious discordances are pointed out here. The correlation phenogram of the cismontane California populations of *magna* is nearly identical to that of the females with two exceptions. First, the two Kern County samples (MNKR, MSKR) clearly belong with the southern cluster of localities in the males. Second, the Los Angeles County sample (OTOP), which clusters with the southern localities in the females, appears with male samples from Arizona and transmontane California. This last cluster includes the Coconino County, Arizona, locality (RCOC) in the males and is more distinct from *sulcata* than in the females. The male Yavapai County sample (SYAV) is again separated from the other *sulcata* OTUs, clustering with the Colorado Plateau locality of *punctata,* which unquestionably belongs to the Californica group of species.

In the Californica species group the similarities between male and female phenograms are more striking than in the Magna group. Other than the unexpected placement of the Colorado Plateau sample of *punctata,* the only difference in composition of major clusters is the inclusion of the Baker County, Oregon, sample (CORB, two individuals) in the San Francisco Bay region cluster in the males. There are also a number of minor differences in the positions of various samples in the large cluster of California and Pacific Northwest localities.

Comparison of the distance and correlation phenograms for males reveals discrepancies which parallel the differences in phenograms of females. In the distance phenogram OLAN, MNYO, and MJSH cluster with the southern cismontane samples. The relatively isolated pair of OTUs, SYAV and RETC, are even more isolated; evidently their unpredictable position is primarily due to the two sulcation and color characters, since results using only the other 19 characters include SYAV with the other *sulcata* samples and place RETC in the Californica species group. Examination of the original similarity matrix based on 21 characters shows that the highest similarity of RETC is to SYAV, but its next six highest similarities are to members of *punctata.* The OTUs RUGU (Modoc Plateau) and CLUC (Santa Lucia Mountains) are more isolated in the distance phenogram in both sexes; in the males CLUC clusters with the San Francisco Bay region samples, in the females with the larger branch of northern California and Pacific Northwest samples. In both sexes the relative isolation of the *punctata* samples is reduced in the distance phenograms, while that of the southern California samples (CLAN, CJAC, CSBR) is increased, but the clusters remain

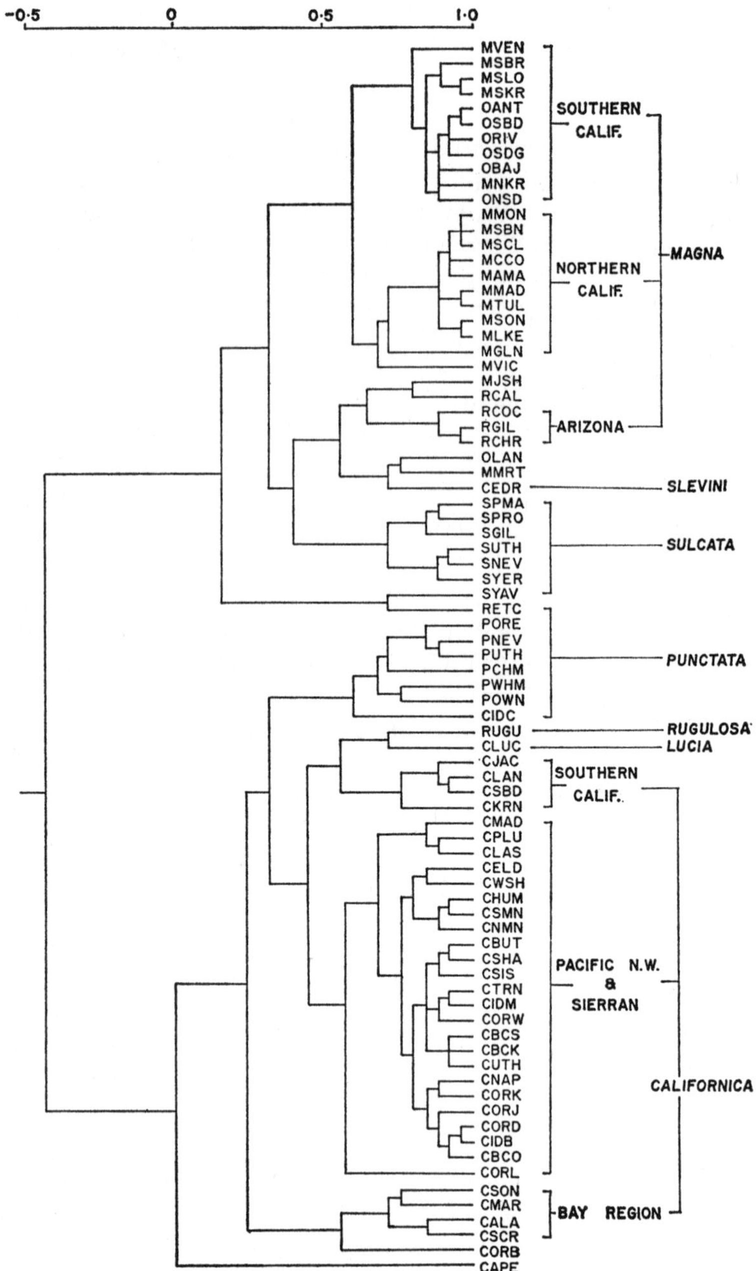

Fig. 40. Correlation phenogram of male samples. See figure 38 for further explanation.

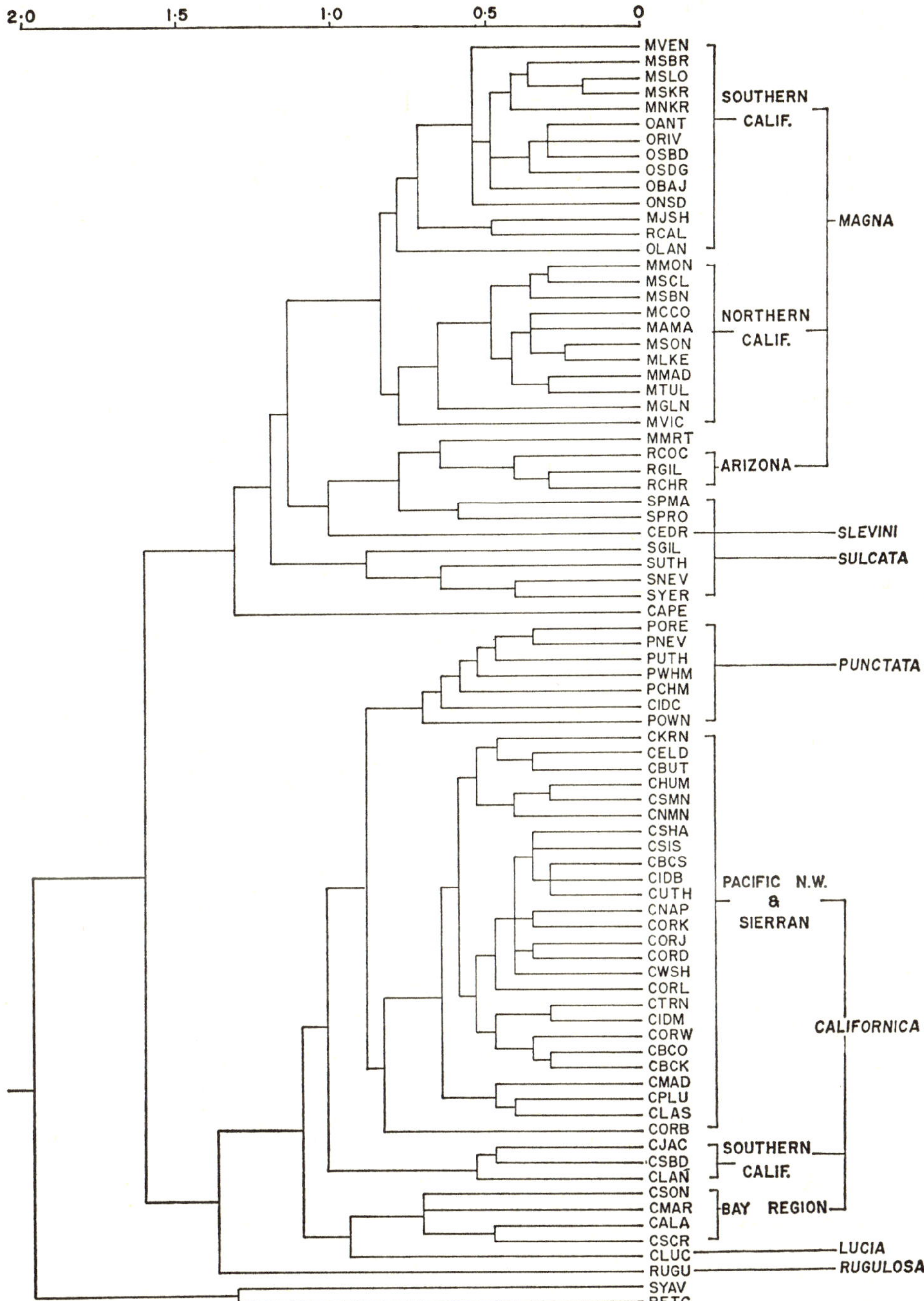

Fig. 41. Taxonomic distance phenogram of male samples. See figure 38 for further explanation.

intact. A confusing aspect of the distance phenogram for males is the inexplicable exclusion of SPMA and SPRO from the other *sulcata* samples. It is also noteworthy that the Baker County, Oregon, sample (CORB) joins the California–Pacific Northwest cluster where it logically belongs.

PHENOGRAMS (DISCUSSION)

Phenograms provide an easily visualized means of representing much of the structure of similarity matrices. In the present study most of the OTUs clearly belong to clusters which have some measure of geographic as well as phenetic integrity, and the membership of analogous clusters in the various phenograms frequently corresponds to the groupings of populations suggested by comparison with the STP diagrams and character-distribution maps. However, the fact that phenograms are often an inadequate means of displaying taxonomic relationships has become generally accepted (Michener, 1970). The problems of interpreting phenograms and detecting the distortion that attends various methods have been the subject of a number of investigations. The effects of using different coefficients of similarity are discussed by Eades (1965), Fisher and Rohlf (1969), Gower (1967), Hall (1969), Minkoff (1965), and Sokal and Michener (1967). These workers and others investigating the use of independent sets of data (Michener and Sokal, 1966), correlated characters (Rohlf, 1967) as well as other factors such as exemplar size, number of characters, standardization of characters, and relative sizes of clusters (Crovello, 1968*a*, 1969; Moss, 1968; Moss and Webster, 1969) have shown that the nonspecificity hypothesis and several of the other assumptions implicit to numerical taxonomy are met to only a limited extent. The most recent comprehensive account of numerical taxonomy is that of Sneath and Sokal (1973). Michener (1970), Sneath (1969), and Moss and Hendrickson (1973), summarize many of the recent trends in the field. Jardine and Sibson (1971) investigate the mathematical properties and appropriateness of various coefficients of similarity and clustering methods. They are critical of many of the methods in current use, and suggest alternative procedures in several cases. Discussion here is limited to a few problems which are highlighted by the present investigation.

1) Perhaps the most serious disadvantage of phenograms is the imposition of a hierarchical structure regardless of the innate structure of the similarity matrix. Hierarchical cluster analysis of random data sets produces phenograms which appear normal in most respects but tend to show very little separation between clustering levels (Rohlf and Fisher, 1968). The cophenetic correlation coefficients are as high as those for many phenograms of noncontrived data and there is no way to detect the randomness by examining the phenograms. This emphasizes the necessity for analyzing the distribution of values in the similarity matrix before accepting the results of cluster analysis. Since biological nomenclature demands a hierarchical structure, phenograms are a suitable means of representation for much taxonomic work, particularly that of a revisionary nature. However, means for detecting the inherent distortion are imperative if phenograms are to avoid the problems of classical methods. Hall (1969) shows that distortion inevitably results from using any of the common coefficients of similarity, regardless of the

clustering procedure employed. He advocates an "average member linkage" clustering algorithm as a means of avoiding distortion. Rohlf (1970) suggests developing adaptive clustering algorithms which would detect trends in the data as the analysis is performed and adjust the hierarchy to these trends. As yet neither of these methods has been tried by others.

At present the commonest measure of agreement between the phenogram and the similarity matrix is the cophenetic correlation coefficient. Most reported coefficients fall in the range 0.70 to 0.90, regardless of the combination of similarity coefficient and clustering technique used, and treatment of a given data set by different numerical methods may produce large differences in placement of OTUs without reflecting this in the cophenetic correlation coefficients (Crovello, 1968a; Moss, 1967). Furthermore, Farris (1969) shows that the cophenetic coefficient does not directly measure the degree to which the phenogram describes the distribution of character states. He also demonstrates algebraically that the unweighted pair group method (UPGA) should yield the highest cophenetic coefficient, regardless of the coefficient of similarity used. The data in the present study were analyzed by applying unweighted (UPGA) and weighted (WPGA) pair-group, single and complete linkage clustering criteria to product-moment correlation and taxonomic distance matrices. At least one major change in cluster membership resulted from each combination of the techniques (only UPGA correlation and distance phenograms are illustrated), yet the cophenetic coefficients were uniformly high. As predicted by Farris, the unweighted pair group method yielded the highest cophenetic coefficient (0.925; males, correlation coefficient). The single-linkage phenogram using taxonomic distance (not illustrated) had the lowest cophenetic coefficient (0.850). It is incorrect to interpret this lack of discrimination simply as an indication of the insensitivity of the cophenetic correlation coefficient to changes in the phenogram, but it shows that these hierarchical clustering schemes portray different but coordinate aspects of the phenetic relationships. One solution to this difficulty is to present a complete set of phenograms, or more typically only those based on single and average linkage relationships. In this study a different approach is utilized to display single-linkage relationships, as elaborated below in the section entitled "Phenomaps." In addition a non-hierarchical clustering technique is explored in the section on taxometric maps.

2) While the problem of measuring distortion in phenograms is important, the specific types of misconstruction which are likely to occur are of greater interest to the practicing systematist. Analysis of relationships in *Coelocnemis* focuses on two types of variation that are prone to distortion in phenograms. The first of these involves the increasing unreliability of the linkage relationships at relatively low levels of similarity, a well-known property of hierarchical clustering (Rohlf, 1968; Michener, 1970). This problem is most severe when some OTUs are phenetically isolated with no high similarities to other OTUs (Moss and Webster, 1969; Moss, 1967). In the phenograms shown here MMRT and CAPE belong to this category and are unpredictably linked with different clusters, even in different species groups. Examination of the similarity matrices reveals that the highest similarities of MMRT are consistently with the Magna species group OTUs and to CAPE. The latter has very low similarities with both the Magna and

Californica groups, explaining its erratic position in the phenograms but the linkage of both MMRT and CAPE at the base of the Californica group clusters (PM, females) is extremely misleading. A similar problem involves the localities SYAV (*sulcata*, Yavapai County, Arizona) and RETC (*punctata*, Colorado Plateau).

A second type of relationship which is inadequately depicted by phenograms occurs when characters vary gradually along a geographical gradient or cline. It was mentioned above that the clinal structure of various series of localities was distorted because at each level of clustering more than one linkage may be determined. This allows several initial cluster nuclei (pairs of very similar OTUs) to grow into clusters of localities, often from opposite ends of the cline. Eventually these clusters are linked, but at a deceptively low similarity level, and in the taxonomic distance phenograms some OTUs outside the cline were admitted before the clusters of the cline were united. The only suggestion of a cline is the fact that some of the OTUs appear with alternate clusters of the cline in different phenograms. For instance MSLO (San Luis Obispo County, California, Magna group, males) clusters with the southern group of localities in the taxonomic distance phenogram but with the northern localities in the correlation phenogram. Univariate analysis showed strongly concordant clines in six characters among these localities, with little significant variation in most other characters, and this pattern is also apparent in ranked lists of highest similarities for each OTU. The highest similarities are consistently between pairs of OTUs in geographic proximity and frequently between geographically contiguous pairs. It is significant that single-linkage phenograms (not illustrated) include all the cismontane localities of *magna* in a single cluster, but still give no indication of the clinal structure.

A final point to be mentioned in regard to the phenograms shown here is their inadequacy for the purpose of making taxonomic decisions. One method for delimiting taxa on phenograms is to place phenon lines at selected levels of similarity. The terminal clusters which are intersected by these lines are accepted as taxa. If this procedure is applied to the phenograms of *Coelocnemis,* a different complement of taxa is obtained for each combination of clustering technique and coefficient of similarity. For instance in the correlation phenograms for females (Magna species group) a phenon line might be drawn at the 0.50 level. This would yield a large cluster of cismontane California localities and smaller clusters consisting of (1) transmontane California–Arizona and (2) *sulcata* localities and RCOC. A number of localities would appear as isolated OTUs (CATA, MSKR, MMRT, CAPE). It is impossible to delimit the same clusters by placing a straight phenon line on the distance phenogram. Any line which leaves all the cismontane California localities in a single cluster will also include some of the transmontane localities (MNYO, MJSH). The localities MMRT, CAPE, and CATA (depending on exactly where the line is drawn) will form a small independent subcluster or part of a larger cluster including the Arizona localities. Similar problems attend the partitioning of the Californica group localities, particularly in respect to RETC and RUGU. Both these OTUs are relatively isolated in the distance phenogram, but join larger clusters in the correlation phenogram. The most acute difficulty concerns the Yavapai County sample of *sulcata* (SYAV), which appears in the *sulcata* cluster in the correlation phenogram but is isolated at the base of the distance phenogram. Further comparison of the phenograms of either sex will reveal many additional inconsistencies.

Division into taxa might also be achieved by distinguishing clusters on the basis of their relative isolation from other related clusters, but this necessitates nomenclatural recognition of such small isolated clusters as SYAV-RETC, which are known to be unnatural on other grounds. This method of partitioning would also be hindered by the different degrees of isolation of many clusters in different phenograms (the *sulcata* localities and those of the southern California mountains of the Californica group are examples).

These problems have been recognized by other workers (Michener, 1970). Probably the difficulty of delimiting taxa is the most serious failing of numerical methods as far as the practicing systematist is concerned. Alternative techniques for partitioning similarity matrices may yield more practical results, but at present the subjective judgment of the individual systematist seems to be as valid as numerical procedures for the purpose of taxonomic discrimination.

TAXOMETRIC MAPS

The difficulties discussed above are by no means novel. The disadvantages of phenograms have received lengthy discussion in the literature and much recent effort in numerical taxonomy has involved the development of alternative methods of representing the information contained in similarity matrices (a few examples include: Brown, 1971; Crovello, 1968*b, c;* Eades, 1970; Farris, Kluge, and Eckardt, 1970; Flake and Turner, 1968; Oxnard and Neeley, 1969). Carmichael and Sneath (1969) provide a summary of the modes of condensation employed in the common methods of cluster analysis and briefly mention the distortion which these entail, then describe a completely nonhierarchical clustering algorithm called "Taxometric Mapping" developed by Carmichael and coworkers (see Carmichael, Julius, and Martin, 1965; Carmichael, George, and Julius, 1968). This technique combines single and average linkage criteria to form clusters of OTUs whose similarities are all above a critical level which is determined by the distribution of similarity values in the study. The clusters are represented by circles whose diameter is proportional to the lowest similarity in the cluster; clusters are separated by center-to-center distances proportional to the sum of their radii plus the distance between the nearest neighbors in the respective clusters. Single OTUs are represented as points. The taxometric maps shown here were constructed from unweighted data.

This nonhierarchical technique is of particular interest in the present study because of the unsuitability of phenograms for representing some of the relationships indicated by univariate analysis. Results for females are presented in figure 42. The most striking aspect of these diagrams is their uncomplicated appearance compared to the phenograms. Nearly all the Californica group localities are included in a single large cluster, with the Colorado Plateau and Modoc Plateau samples isolated as satellite OTUs. The Magna species group consists primarily

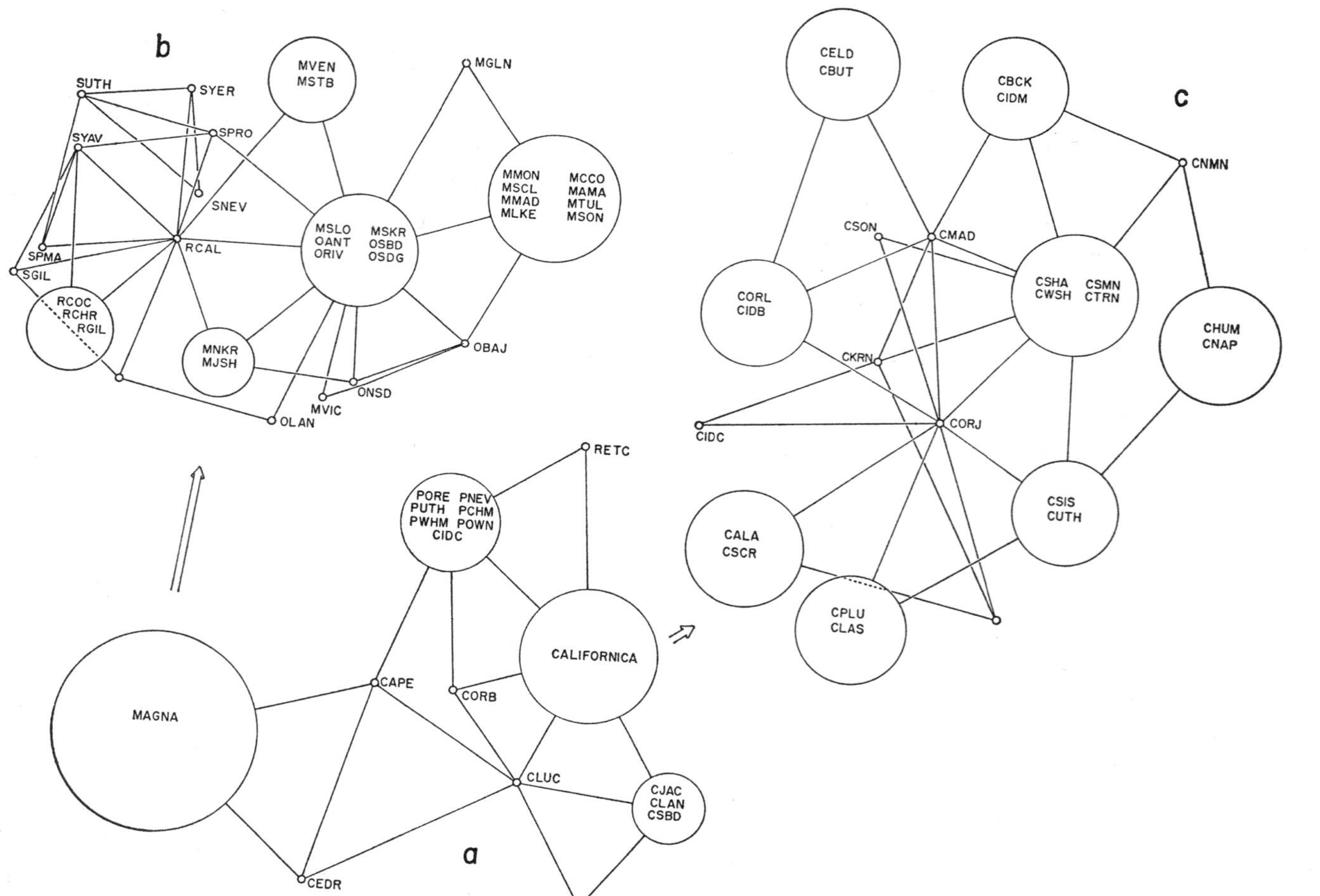

Fig. 42 Taxometric map for female samples. (a) Clusters generated from data representing 75 OTUs. (b) Clusters generated by subjecting the OTUs of the large *magna* cluster in (a) to a second cycle of clustering. (c) Clusters generated by subjecting the OTUs of the large *californica* cluster in (a) to a second cycle of clustering.

of one very large cluster containing all the cismontane California localities, some of the transmontane localities and most of the *sulcata* samples. Two of the montane Arizona localities form a second cluster, and the remaining OTUs are isolated.

Greater detail can be resolved in large clusters by subjecting them to further clustering cycles with the other OTUs removed. The results of a second round of clustering of the large groups of *magna* and *californica* localities are shown by the detail indicated by arrows. The *magna* localities are partitioned into several 2- or 3-member clusters of geographically proximal OTUs and a large number of single-member clusters. Significantly there is no suggestion of a north-south bisection of the cismontane California localities, as indicated by the phenograms. The *californica* cluster subdivides into a large group of Pacific Northwest localities and smaller subclusters of the southern California and Sierran samples: two other small subclusters include *punctata* samples. Again the number of isolated OTUs is large.

The results for males (fig. 43) emphasize the phenetic gaps in the Californica species group and minimize those in the Magna group. The localities of the *californica* group segregate into a large cluster of California–Pacific Northwest localities, and two smaller clusters representing the Great Basin and montane southern California. Both the latter are quite close to the large California–Pacific Northwest cluster. The localities RETC (Colorado Plateau), CORB (Baker County, Oregon), RUGU (Modoc Plateau), and CLUC (Santa Lucia Mountains) are isolated. The entire Magna species group is contained in a single large cluster, with the exception of CEDR (Cedros Island) and CAPE (San Jose del Cabo), which are isolated.

The results of a second cycle of clustering of the males is indicated by arrows. The cismontane localities of *magna* split into a pair of large subclusters; smaller subclusters contain the Arizona samples and two of the transmontane California localities. The OTUs MVEN and MSBR are inexplicably isolated in yet another small subcluster. The Californica species group is subdivided into numerous small subclusters that mostly contain geographically proximal localities. In both cases the number of isolated OTUs is large.

It is clear from these results that taxometric maps do not provide a completely acceptable alternative to phenograms and other better established methods of representing numerical taxonomic results. Since the clusters are located in hyperspace there is no satisfactory way of illustrating relationships of more than three dimensions. This is not a serious drawback when the number of clusters is small, but, with more than about a dozen, intercluster relationships are severely distorted and the map becomes unmanageably complex. It would also be very helpful to know the shapes of the clusters, which can be inferred from the map-cluster analysis to only a limited extent. Another obvious improvement would be the inclusion of an algorithm to minimize the sum of intercluster distances while preserving the centrality-peripherality relationships of the clusters. Eades (1970) points out that taxometric map results depend on the cluster which is selected as the starting point, which is probably the most serious criticism of this technique. An unexpected feature of the taxometric maps is the isolation of a rather large number of samples in single-member clusters. Most of these are rather isolated in the phenograms as well, but RCOC, SGIL, SUTH, and CALA in particular

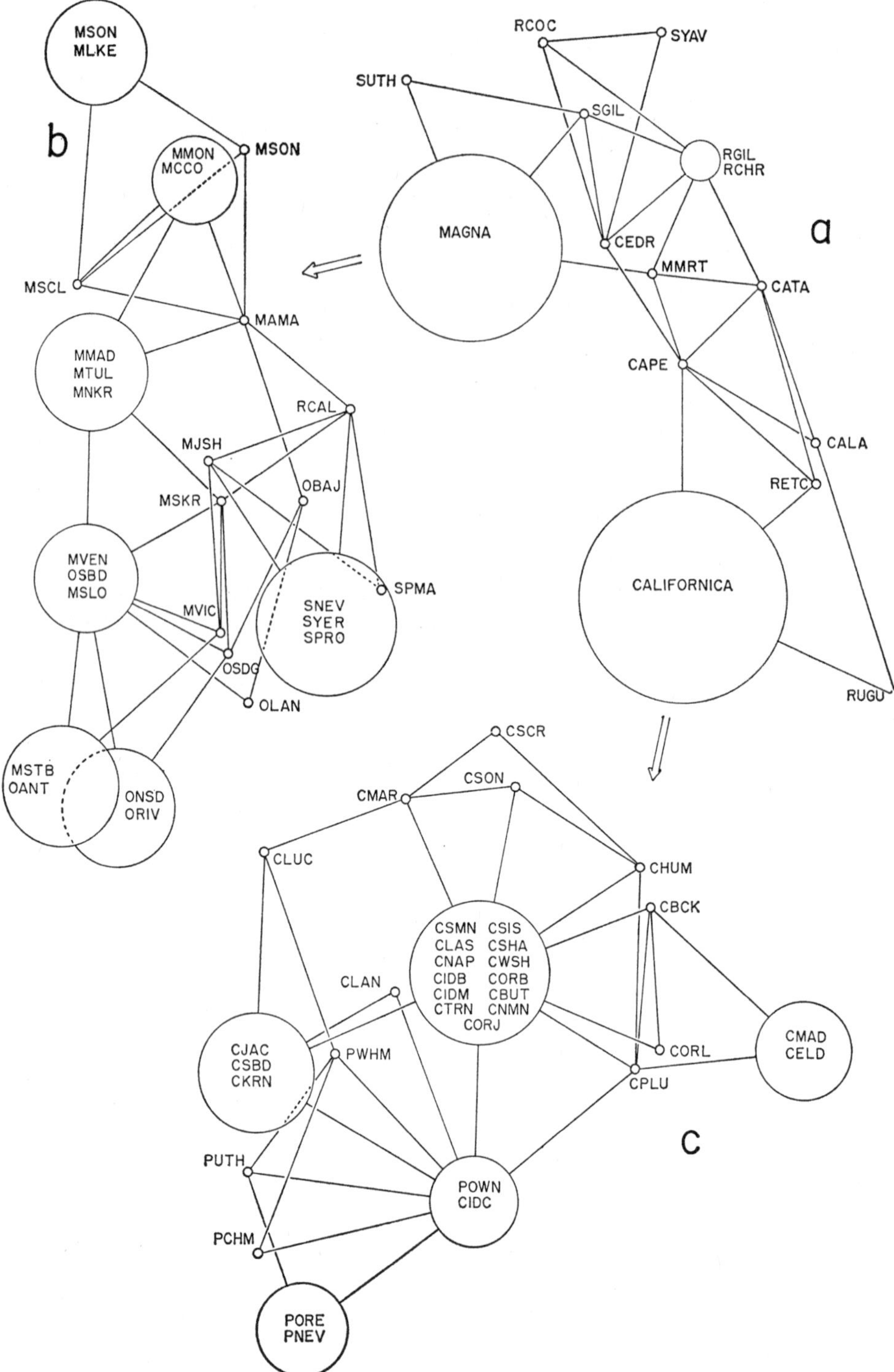

Fig. 43. Taxometric map for male samples. (a) Clusters generated from data representing 75 OTUs. (b) Clusters generated by subjecting the OTUs of the large *magna* cluster to a second cycle of clustering. (c) Clusters generated by subjecting the OTUs of the large *californica* cluster to a second cycle of clustering.

showed clear membership in larger clusters in the phenograms. On the other hand, the locality MSKR is isolated in the phenograms but clearly belongs in the large Magna group cluster according to the taxometric map.

Aside from these difficulties taxometric maps of the genus *Coelocnemis* reveal certain aspects of phenetic structure which are concealed by phenograms. The nonhierarchical clusters of the taxometric maps minimize the differences between the geographical segregates within each species group, although subsequent clustering cycles verify the presence of these groups in most cases.

The greatest advantage of nonhierarchical clustering seems to be its more realistic treatment of gradual variation; the diverse series of cismontane Magna group populations which are consistently separated in the phenograms appear as a single cluster in the taxometric maps; similarly the San Francisco Bay region localities of *C. californica* are not separated by a single cycle of clustering. The fact that each species group is subdivided in only one sex suggests that these divisions are not as important as the phenograms indicate. Taxonomic decisions based on the taxometric maps would obviously be quite different from those derived from the phenograms, but no less subjective. Taxonomic recognition of the relatively large number of isolated OTUs would be particularly undesirable, since most show close relationships to larger clusters in all other methods of analysis.

PHENOMAPS

Much of the ambiguity in the results of the various methods of cluster analysis, especially the hierarchical procedures employed in constructing phenograms, derives from distortion at low levels of similarity. This problem is particularly acute when average linkage criteria are employed, but single-linkage phenograms as well as other types of analysis have the additional disadvantage that the geographic and phenetic relationships of the OTUs cannot be viewed simultaneously. Both these problems can be partially overcome by plotting the OTUs geographically on a map and indicating the highest similarity relationships by connecting lines. The resulting diagrams show both geographic and phenetic relationships among OTUs, suggesting the name "phenomaps," which is employed here. These diagrams are well suited for displaying relationships between OTUs which share geographic as well as phenetic proximity, exemplified by the numerous samples in the present study, but could also be valuable in analyzing zoogeographic relationships between taxa widely separated geographically. In the diagrams presented here the two highest similarities are shown for each OTU. A ranked list of all the similarity values generated from each matrix was truncated by deleting the top 5 percent of the values; for each OTU the similarity relationships falling in or above the highest decile of this truncated list are represented by solid lines. Similarity values below this level are represented by dashed lines. All OTUs are included except CAPE, which shows inconclusive similarities to both species groups. "Phenomaps" are illustrated for taxonomic distance; correlation-coefficient diagrams are very similar for both sexes. To improve clarity, each species group is presented separately, since the highest similarity relationships cross species-group boundaries in only a few cases mentioned below. An alternative approach might involve showing the linkages between all geographically contiguous pairs of OTUs (see Gabriel and Sokal, 1969, for an operational definition of contiguity).

Related graphic techniques for analyzing phenetic relationships include the minimal connection networks of Prim (1957) and the graphs and contour diagrams of Moss (1967) and Hubac (1964). Prim-networks were constructed for *Coelocnemis*, but have the disadvantage of not showing the degree of interconnectedness within clusters. The methods devised by Moss and Hubac entail plotting the position of each OTU by triangulation; using first and second order relationships, without regard to geographic location. In this respect they are similar to the taxometric mapping procedure of Carmichael and Sneath (1969). More sophisticated methods of illustrating geographic and phenetic relationships, not discussed here, include the construction of contour maps and trend surfaces (James, 1970); Adams, 1970; Pauken and Metter, 1971; Marcus and Vandermeer, 1966; Sneath, 1967; Chorley and Haggett, 1968) and multidimensional scaling (Kruskal, 1964; Holloway and Jardine, 1968).

Phenomap relationships of female samples of the Magna species group are shown in figure 44. The cismontane California populations obviously form a tightly knit group with almost no linkages to outside OTUs. More significant, however, is the propensity of nearest phenetic neighbors to be geographically contiguous or in close proximity. This is particularly evident in the central Coast Ranges where the clinal change in characters is most pronounced. The relationships in southern California are more diffuse, reflecting the relatively uniform phenotype throughout this area, and the southernmost samples (OSDG and OBAJ) are linked to samples in the central Coast and Transverse Ranges, recall-similar patterns of variation revealed in several characters by univariate analysis. There is a gap between MNKR and MTUL in the southern Sierra Nevada, but the fact that the third highest similarity for both MNKR and MTUL is with the other member of this pair reduces the significance of this disjunction. While linkage relationships are linear along the central Coast Ranges and Sierra Nevada, with none crossing the central valley, they are less clear-cut in the north, especially north of San Francisco Bay. The relationships of the three North-Bay samples are entirely with one another or with the Sierran samples, and similarities are of a relatively low level. The high similarity between MCCO (Contra Costa County) and MAMA (Amador County) is surprising. The tendency of linkages to cross the Sacramento Valley is possibly explained by the fact that strips of riparian woodland originally criss-crossed the valley and are still present today in reduced extent, whereas the more arid, nearly treeless San Joaquin Valley to the south has probably been a strong barrier to gene flow for a much longer time. (During the pluvial portions of the Pleistocene and even in historical times the southern San Joaquin Valley has supported large areas of marshland. These would also be a strong barrier to dispersal to *Coelocnemis*.) It is interesting to point out here that the central coast race of the salamander, *Ensatina eschscholtzi*, occurs in the Sierra Nevada opposite the Sacramento delta (Stebbins, 1949). During the last glacial maximum the delta region was apparently sufficiently mesic to allow dispersal between the Coast Ranges and the Sierra.

A second axis of high similarities unites the three Arizona localities of *C. magna*,

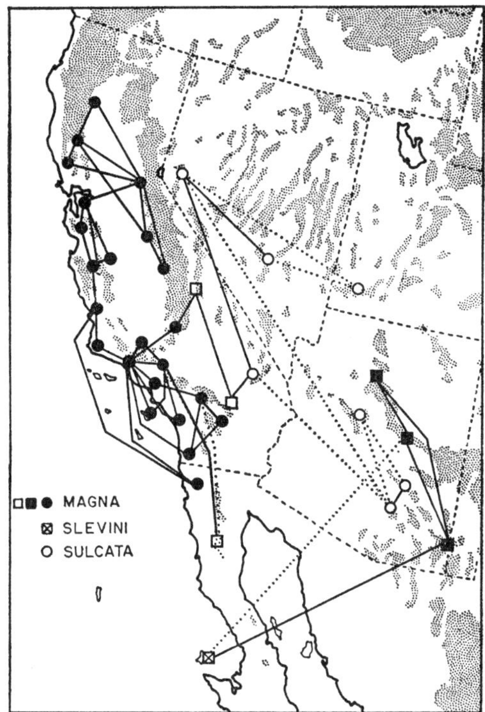

Fig. 44. Phenomap of female samples of the Magna species group. Symbols represent sample localities. Connecting lines represent the two highest similarities for each sample. Solid lines indicate very high values; dotted lines indicate lower values. See text for further explanation.

although the Coconino County sample is relatively isolated and shows secondary relationships to the transmontane California localities. The Cedros Island population is most similar to those in southern Arizona rather than to the geographically closer populations in mainland Baja California. The transmontane populations of the desert mountain ranges show predominantly low similarities to both the Arizona and California samples and to one another. The *sulcata* localities are linked only among themselves, but at a uniformly low level. The Great Basin and Arizona segments of *sulcata* tend to remain disjunct, but there are several interconnections; the OTU SPRO (Providence Mountains) also shows a linkage to the Little San Bernardino Mountains population (MJSH). The low similarities among the transmontane and Great Basin samples probably stem from the isolation of these populations in the basin ranges, although limited gene flow is indicated by a few collection records from the intervening playas.

The phenomap linkages among the male samples of the Magna species group (fig. 45) are highly similar to those of the females. The trough of low values north of the Transverse Ranges of California is more pronounced in the males, but pairs of localities on opposite sides of this disjunction are again linked by relatively high values. The linkages crossing the Sacramento Valley reinforce the suggestion of past gene flow and the absence of linkages between the central Sierra and Coast Ranges emphasize the effectiveness of the San Joaquin Valley as a bar-

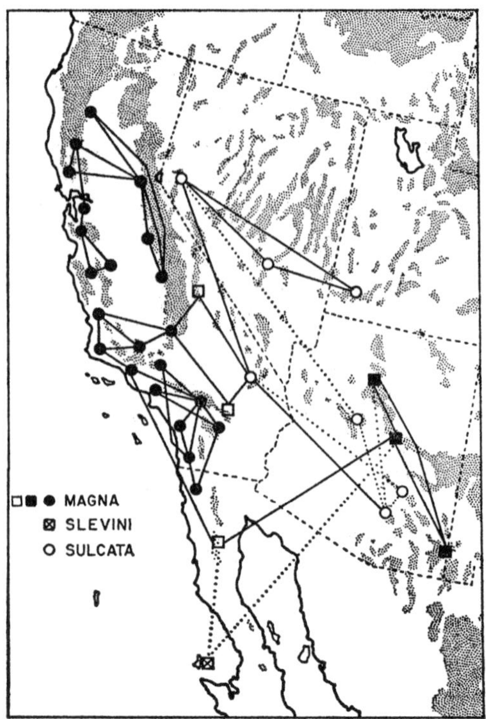

Fig. 45 Phenomap of male samples of the Magna species group. See figure 44 for explanation.

rier. The transmontane California and Cedros Island samples again show ambiguous similarities to those of both cismontane California and Arizona, as well as among themselves. The only instance in which nearest phenetic neighbors are members of different species groups involves SYAV (Yavapai County, Arizona) of the Magna group and RETC (Colorado Plateau) of the Californica group, but the next six highest relationships of SYAV are with *sulcata* samples, the next six of RETC with *punctata* localities. The phenomap relationships of the Arizona populations of *sulcata* are rather confused, but the Great Basin populations form an interconnected cluster. As in the females, SPRO shows high similarities to a variety of samples.

The phenomap of the Californica species-group is illustrated in figures 46 and 47. While the San Francisco Bay region, the southern California, and the Sierran samples tend to be most similar to geographically proximal samples, the large group of Pacific Northwest and northern California populations are frequently separated from their nearest phenetic neighbors by great distances or major physical barriers. Since lack of geographic order is logically expected among phenetically homogeneous OTUs, these results agree with the phenograms presented earlier. Contrastingly, the linear north-south relationships along the Sierran chain underline the phenetic integrity of these populations. Linkages in the San Francisco Bay area are predominantly between localities within this region in the males, but predominantly to the north in the females, and in both sexes the similarities

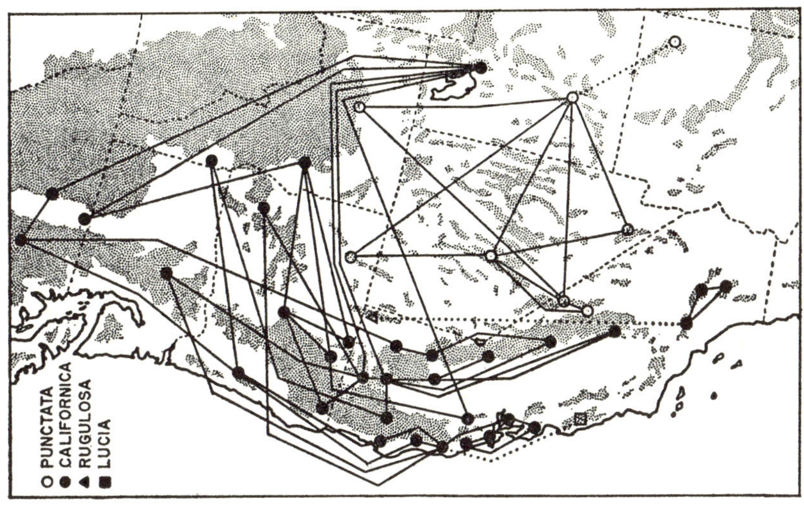

Fig. 47. Phenomap of male samples of the Californica species group. See figure 44 for explanation.

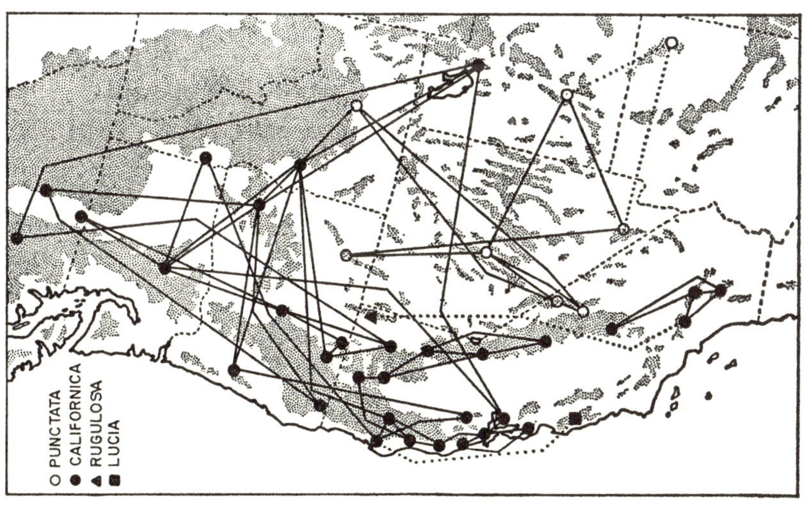

Fig. 46. Phenomap of female samples of the Californica species group. See figure 44 for explanation.

are relatively low. The unexpected linkages of CALA (Alameda County) females to the Utah and Washington samples further confuse the relationships among this group of populations. It should be recalled that the distribution of *C. californica* is discontinuous in the region south of the Russian River, suggesting the possibility of differential evolution among the various populations. The fact that these populations differ from those to the north in only a few cuticular sculpturing characters which vary along a steep cline, may also explain the low similarities in the San Francisco Bay region. The anomalous Alameda County sample contained several individuals with extreme values for the sculpturing characters, possibly accounting for its otherwise inexplicable linkage relationships. Regardless of the confusing relationships in this region it is evident from the nearest-neighbor diagrams that the bay region samples are not nearly so isolated as the phenograms imply.

Aside from their high similarities to CKRN (Kern County) females, the southern California populations are completely interlinked; their next highest similarities are to Sierran populations, but drop off sharply. The two OTUs CLUC (Santa Lucia Mountains) and RUGU (Modoc Plateau) are isolated in the nearest-neighbor diagrams, as expected from other methods of analysis.

The configuration of the nearest-neighbor relationships of the *punctata* samples resembles that of *sulcata,* the other Great Basin species, but most of the similarities are of a higher level. There is only a single linkage to a population outside the Great Basin (PIDA to CIDB). The axis of linkages between PIDA (Craters of the Moon, Idaho) and the California populations (POWN, PWHM) recalls the phenetic intermediacy of all these OTUs.

BIOGEOGRAPHIC RELATIONSHIPS

Distributional patterns of organisms normally conform to climatic or other ecologically important factors. By plotting the occurrence of critical climatic or edaphic parameters it is sometimes possible to deduce the range of taxa and, conversely, to use certain species as climatic indicators. For example, the maritime species of closed-cone pines (*Pinus muricata* D. Don., *P. radiata* D. Don., *P. remorata* Mason) of the Pacific Coast occur only along the narrow coastal strip subject to summer fog and mild temperatures and on some of the California islands. However, existing distributions reflect past climates as well as present ones, so that it is not usually possible to explain present ranges of organisms merely by their ecological requirements. As an illustration, the closed-cone pines mentioned above occur discontinuously along the coast and are absent from many apparently suitable sites. Axelrod (1967) convincingly explains the present distribution as the result of the hot, dry climate of the Xerothermic period (ca. 8000–3000 B.P.) which disrupted a continuous coastal forest (see also, Axelrod, 1966). That the original distribution was much more extensive is unequivocally demonstrated by Miocene to late Pleistocene fossil records from coastal areas in which the pines do not now occur.

Biogeographic conclusions about organisms with no known fossil record, such as the species of *Coelocnemis,* must be much more tenuous, but by utilizing the growing amount of paleoecological information it is possible to reconstruct some of the past distributional changes which occurred. A basic assumption which underlies such an interpretation is that the ecological requirements of the species have remained relatively constant and that the present distributional patterns are a result of attenuations or expansions of range in response to climatic changes. This is indicated by the highly discontinuous present distribution of the genus *Coelocnemis* and by the fact that the beetles are all flightless, severely limiting their dispersal powers. All species of *Coelocnemis* are brachypterous. The elytra interlock securely with one another and the abdominal sterna, and the flight muscles are atrophied, indicating that the apterous condition is primitive in the genus. Obviously it would have been impossible for a flightless species to colonize the isolated desert mountain ranges and Baja California localities unless more mesic habitats once connected them to the woodland and chaparral associations in which the beetles now occur.

Most of the distributional complexities of *Coelocnemis* probably originated during the Pleistocene (as detailed below), but the beetles may have invaded southern Baja California during the Pliocene when the peninsula assumed essentially its present outline. Accordingly, the following discussion considers the climates of both late Tertiary and Quaternary times. The climates and composition of the floras and faunas of this time span have been the subject of intensive recent investigations. The pioneering work of Axelrod and his associates is summarized by Axelrod (1958*a*, *b*, 1967) and Darrow (1961). Much of the information on Pleistocene climates in western North America is the result of pollen analysis (e.g., Heusser, 1960; Martin and Mehringer, 1965; Mehringer, 1965; Baker, 1970). The macrofossil investigations of Wells (1970), Wells and Jorgenson (1964), and Wells and Berger (1967) seem to substantiate a modified version of the palynological studies. Many aspects of Pleistocene biology and geology which are not discussed here are comprehensively reviewed in the papers by Flint (1957), Hubbs (1958), Wright and Frey (1965), and Cushing and Wright (1967), which contain diverse discussions of the paleoecology of western North America, including purely biogeographic works. Regional studies of the Baja California peninsula (Beltran, Garth, and Savage, 1960) and the California islands (Philbrick, 1967) include paleoecological information pertinent to the present study. The work of Coope (1967, 1970) using fossil beetles to characterize Pleistocene climates in Europe has only limited bearing on problems in western North America but is important in demonstrating that insects as well as plants may be used as paleoclimatic indicators. The lucid general account of the geological history of western North America by King (1958) contains many references to more specific regional and local works of interest to both geologists and biologists.

While it is not the purpose of this paper to provide a lengthy analysis of theories for the origins of the vegetation of western North America, it is necessary to mention a major divergence in emphasis among modern workers. Axelrod's concept that major vegetational assemblages arose in Miocene times or earlier, migrated more or less intact in response to climatic and orogenic changes and surived in modified form to recent times is rejected by Johnson (1968), Whittaker (1965), Whittaker and Niering (1968), Mason (1947), and Wolfe (1969) who favor the theory of individualistic adaptations to environmental forces as first set forth by

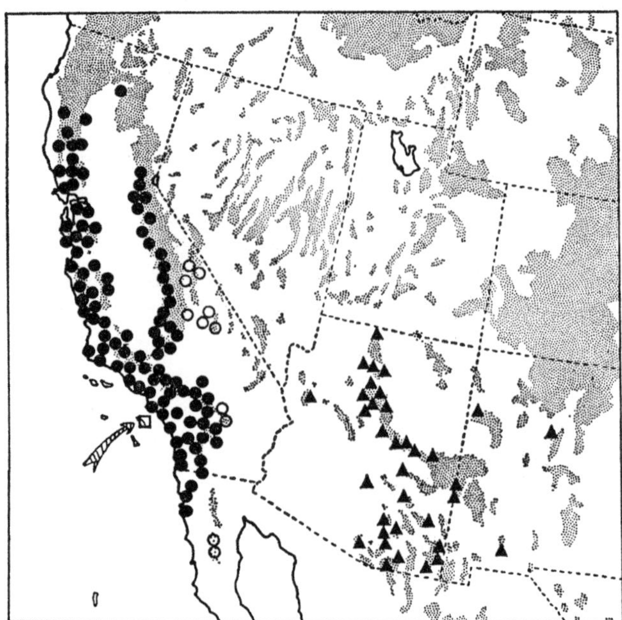

Fig. 48. Distribution of collection localities for *Coelocnemis magna* LeConte. Solid circles represent the cismontane California series. Open circles represent the Sierra Martir and various transmontane California series. Triangles represent the montane Arizona series. The square represents the Santa Catalina Island series.

Gleason (1926). In the words of Johnson, "plant communities are transient, incidental aggregations of species that share similar ecological tolerances," and Mason concludes that the only consistent species of the redwood association through the Tertiary and Quaternary is the redwood itself. While it is undoubtedly true that the composition of plant communities has changed drastically through geologic time it also seems obvious that some of the changes evidenced by the assemblages of species in fossil floras are indicative of past changes in climate. It is also clear that the distributions of many species of both plants and animals have changed radically since the late Tertiary and that Axelrod's interpretations explain many of the present distributional phenomena. Consequently, I have followed the example of Peabody and Savage (1958) in accepting Axelrod's terminology and most of his paleoecological conclusions, modifying them when the available evidence suggests alternatives.

The foregoing morphological, ecological, and biochemical information clearly demonstrates that the beetles of the genus *Coelocnemis* represent two alliances: the Californica and Magna species groups. One of these, with obvious northern affinities, is distributed today throughout the montane and forested coastal regions of western North America and in the Great Basin. This area, called the Vancouveran faunal zone by Van Dyke (1919, 1939), and Linsley (1958) encompasses the Pacific Northwest, coastal and Sierran California, and in a broader sense the western Rockies, the Wasatch Range, and the higher ranges in the Great Basin and Colorado Plateau. The Vancouveran faunal area, typified by ample winter

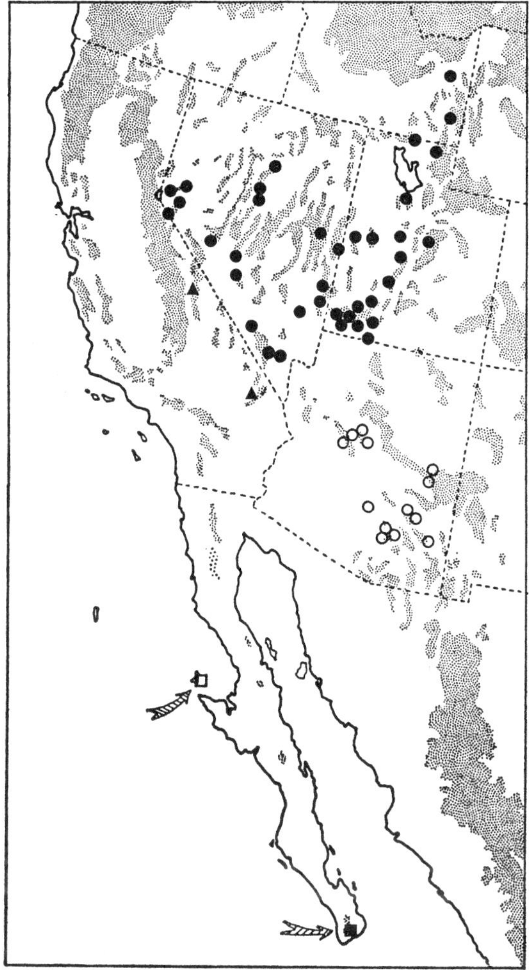

Fig. 49. Distribution of collection localities for *Coelocnemis sulcata* Casey, *C. slevini* Blaisdell and an undescribed species from the Cape region of Baja California. Solid circles represent the Great Basin series of *sulcata;* open circles represent the montane Arizona series; solid triangles represent the Providence Mountains series. The solid square represents the undescribed species. The open square represents *slevini*.

rainfall and moderate annual temperature extremes, is characterized by species derived from the Western Conifer Element of the Arcto-Tertiary Geoflora. The Californica species group of *Coelocnemis* occurs throughout the Vancouveran except in the southern Rocky Mountains (figs. 50–51), a distributional pattern shared by few other Tenebrionidae.

The Magna species group presently occupies the live oak-conifer woodland of cismontane California and northern Baja California (the Californian zone of Linsley) as well as the woodland areas of Arizona and New Mexico, the Great Basin, and southern and north central Baja California (figs. 48 and 49). This distribution corresponds very closely to the present relictual occurrence of the Madro-Tertiary Geoflora. This area contains an extensive tenebrionid fauna

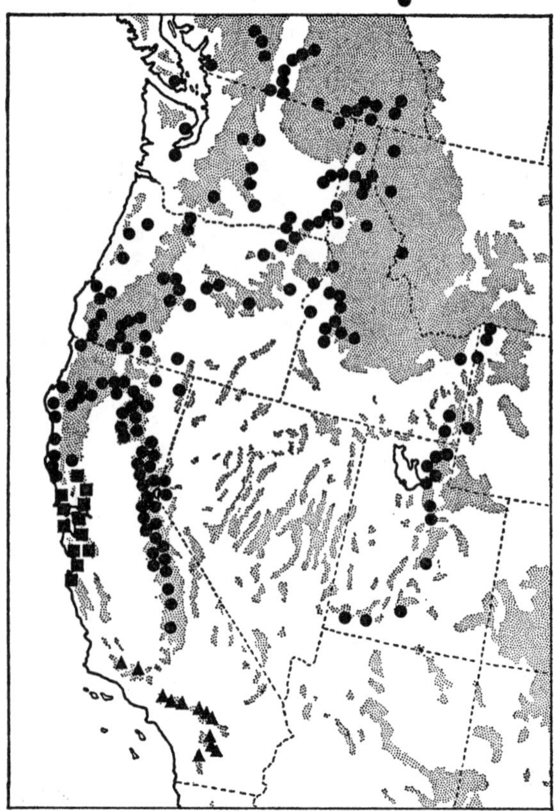

Fig. 50. Distribution of collection localities for *Coelocnemis californica* Mannerheim. Circles represent the Pacific Northwest and Sierra Nevada series. Squares represent the San Francisco Bay region series, and triangles represent the southern California series.

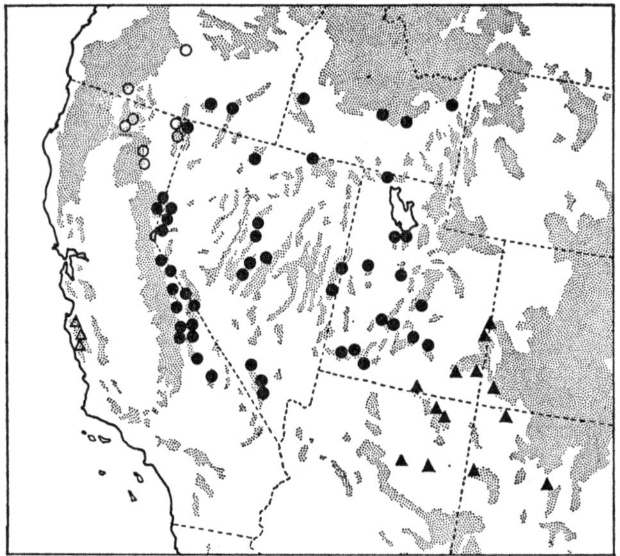

Fig. 51. Distribution of collection localities for *Coelocnemis punctata* LeConte, *C. lucia* n. sp. and *C. rugulosa* n. sp. Solid circles represent the Great Basin series of punctata; solid triangles represent the Colorado Plateau series. Open circles represent *rugulosa*. Open triangles represent *lucia*.

characterized by many endemic genera and species. If Axelrod's geofloral reconstructions are correct it would seem that the species groups of *Coelocnemis* had already diverged by the late Tertiary when the extensive Madro-Tertiary plant associations were disrupted by drier, less temperate climates. There is no concrete evidence indicating whether the Arcto-Tertiary or the Madro-Tertiary species were antecedent in time, but the phenetic similarity of *Coelocnemis* to *Iphthimus* of Holarctic distribution suggests that the two evolved in northern forests as early as the Miocene. The western species of *Iphthimus* are slightly brachypterous today, suggesting that this trend was continued in *Coelocnemis,* which developed a shorter, thicker body adapted to the looser bark of older, more decayed snags. Efficient walking becomes important in flightless species, explaining the relatively long legs in *Coelocnemis* and the convergence in general appearance to the ambulatory genus *Eleodes.* The short-legged, flattened species of *Iphthimus* are adapted to scuttling beneath the bark of recently fallen trees and have difficulty walking on uneven surfaces; the two genera are not commonly found in close association.

The Magna species group may have diverged from the Californica group as early as the Eocene, when populations of Madro-Tertiary plant species first appear in the fossil record (see Axelrod, 1958*b*), more probably during the Miocene or Pliocene, when Madro-Tertiary species achieved widespread distributions. Howden (1963, 1966) postulates a similar age for several genera of scarab beetles, based on distributional evidence.

The only other North American genus which is morphologically similar to *Coelocnemis* is *Oenopion,* presently restricted to Madro-Tertiary oak woodlands in the Sierra Madre Oriental and possibly of southern Texas and New Mexico. *Oenopion* differs from all species of *Coelocnemis* in mouthpart structure and defensive secretion composition, suggesting a separate derivation. On the other hand, the peculiar structure of the mentum, nearly identical larvae, and very similar defensive secretion composition argue against convergent evolution of the Magna and Californica species groups from different parent species. Before discussing the probable geological history of each species group in detail, it should be emphasized that they differ considerably in ecological tolerances and that barriers to distribution of one group may act as corridors for the other.

The Magna Species Group

During the early and middle Miocene much of central Baja California was submerged and during the middle Pliocene climate became drier; the local desert of the region may have expanded at that time (Axelrod, 1958*a*). Johnson (1968) disagrees with Axelrod's plant associations and indicates that the floristic elements composing the southwestern deserts are of diverse age, some of them probably originating during the Miocene or earlier. The extremely diverse tenebrionid fauna of this region, with many highly specialized arenicolous species, also suggests that appreciably large tracts of arid land have existed in the Southwest for a long time, although no actual date can be assigned. At any rate, local areas of the Sonoran Desert have probably been quite arid since the early Pliocene, including the periods of Pleistocene glaciation.

These data and conclusions suggest that the relictual populations of *Coelocnemis* on Cedros Island and in the Cape region of Baja California have been isolated since about the late Miocene to middle Pliocene by barriers of water or aridity. However, recent analysis of paleomagnetic patterns of the seafloor near the mouth of the Gulf of California show that most of the northwestward movement of the peninsula has occurred during the last four million years (Larson et al., 1968). Prior to this time the gulf may have existed only as a very narrow strait, as suggested by Howden (1963). This conflicting evidence indicates that the Baja California populations have been isolated for a much longer period of time than the numerous allopatric races farther north, but the exact time of isolation is uncertain.

Certainly the relatively high degree of morphological differentiation of the Baja California populations from the more uniform beetles of southwestern United States would indicate a relatively long and continuous separation. The cape region beetles (unnamed) are the most divergent of the Magna species group. They are similar in rugosity and some body proportions to members of the Californica species group, suggesting the possibility that they represent a relictual occurrence of a stock which gave rise to both species groups. If this were true, however, the Californica group would be secondarily adapted to Arcto-Tertiary habitats, which seems unlikely on the basis of evidence presented earlier. The Cedros Island beetles (*C. slevini*) are most similar in cuticular sculpturing to the beetles in southern Arizona but share some characters also with southern California populations. The relatively long, narrow mentum distinguishes *C. slevini* as the only taxon of the Magna group with an appreciable structural divergence in the mouthparts, and it is possible that it has been isolated nearly as long as the beetles from the Cape Region. The range of *C. magna* probably extended further south in Baja California during the Pleistocene glaciations, but probably not far beyond the Sierra San Pedro Martir. Macrofossil evidence indcates that the pinyon-juniper zone dropped only about 600 m in the southern Great Basin during the pluvial maximum (Wells and Berger, 1967) and that the woodland was then associated with desert shrubs such as *Ambrosia* and *Atriplex*, which now occur in the area. While the higher divides of the Mojave Desert probably supported arid woodland, there is no implication that the lower basins and playas were much more mesic than today. The degree of morphological differentiation between *C. slevini* and *C. magna* is much greater than that between many populations of *magna* that have probably been isolated in desert mountain ranges only since the pluvial maximum, suggesting again an isolation of considerable duration. In general, the status of the Baja California species of *Coelocnemis* seems to support Howden's contention (1966) that deserts have probably existed as barriers in North America since about Miocene times.

The presently fragmented distribution of the Magna species group in southwestern United States seems to be largely a result of climatic changes in the Pleistocene. While the middle and late Pliocene climate was probably considerably drier and cooler than that of the Miocene (Axelrod, 1967), several typical Madro-Tertiary plants now restricted to cismontane California or montane Arizona are represented in fossil floras from west central Nevada. These floras suggest an arid woodland vegetation, probably ecotonal between the retreating Arcto-Tertiary forests and the invading Madro-Tertiary elements. Axelrod postulates that the desert vegetation of western North America developed its modern aspect after middle Pliocene time, but even if extensive deserts were present before this time as postulated here, it may be assumed that woodland existed on the slopes of the mountains then present in the Basin and Range Province. During the Pliocene many Madro-Tertiary species probably achieved their maximum range in California, extending at least as far north as San Francisco (Axelrod, 1958*a, b,* 1967). During this period *Coelocnemis magna* and *C. sulcata* probably occupied much the same area as at present, possibly with a mosaic distribution in the Great Basin and Mojave Deserts and continuous populations in the more temperate climates of cismontane California and southern Arizona.

During the Plio-Pleistocene transition the climates of western North America were apparently much cooler and more humid than at present. The first glacial stage (Nebraskan) was apparently cool and moist enough to allow yellow pine to migrate into the lowlands and lower slopes of the southern California mountains from the Sierra Nevada. Yellow pine forest also ranged east to the Panamint Valley at this time (Axelrod and Ting, 1960) and the Soboba Flora near San Jacinto shows intermingling of yellow pine, bigcone spruce, and Coulter pine on the floor of the present San Jacinto Valley (Axelrod, 1966). Late Pliocene and early Pleistocene fossil floras are rare, but the evidence presented above suggests disruption of the extensive Madro-Tertiary vegetation of the middle Pliocene. The processes responsible for invasion of the Arcto-Tertiary flora into southern California apparently caused a simultaneous expansion of the woodland habitats bordering the mountains of the Basin and Range Province. It seems likely that fluctuations of this sort accompanied each Pleistocene glaciation, imposing repeated fragmentation and consolidation of both Madro-Tertiary and Arcto-Tertiary distributions. During the Quaternary, populations of *C. magna* and *C. sulcata* now isolated in the Sierra San Pedro Martir, Little San Bernardino, Providence, and Hualapai Mountains must have been connected several times with the more extensive populations in cismontane California and montane Arizona. The last contact between these isolates could have been as recently as 12,000–20,000 years ago, during the Mankato substage of the Wisconsin. The minor extent of morphological divergence of these isolated populations supports the idea of recent contact. The Hualapai Mountains individuals are indistinguishable from montane Arizona beetles, while those in the Little San Bernardino Mountains are extremely similar to other southern California specimens. Beetles from the Sierra San Pedro Martir are more distinct, showing ambiguous phenetic similarities to both southern California and Arizona specimens, but it seems likely that they resulted from a migration from southern California in Plio-Pleistocene or Pleistocene times. A more remote possibility exists that the Martir populations date from the Pliocene, which could explain the similarity to material from Arizona. The populations from the central Owens Valley and associated basins and ranges are practically indistinguishable from specimens from the southern and central Sierra Nevada, suggesting that more continuous populations may have existed in this area during the Pleistocene glacial maxima. The occurrence of *C. magna* in this area is matched by the presence of an endemic species of Alligator lizard (*Gerrhonotus pana-*

mintinus Stebbins) in the Panamint Range (Stebbins, 1958).[4] The lizard, however, is apparently restricted to one or a few moist canyon bottoms, while *Coelocnemis* has been taken in the arid treeless basins as well as in more mesic riparian situations. Records from Walker Pass, Kern County, suggest that very limited gene flow may occur between the Owens Valley and cismontane California populations at present. However, it is probable that they were briefly separated during the Xerothermic Period. The effects of this period are now reflected in the disjunct northerly distribution in California of plants such as *Quercus palmeri* Engelmann, *Adenostomum sparsifolium* Torrey, and other chaparral and coastal sage species (Axelrod, 1958*a*, 1966). A number of semidesert or desert species also extend disjunctly along the inner Coast Ranges as far north as the Mount Hamilton Range (Sharsmith, 1945). The disrupted occurrences of such xerophilous Tenebrionidae as *Edrotes ventricosus* LeConte, *Triorophus subpubescens* Horn, and *Araeoschizus sulcicollis* Horn in the same area (Doyen and Opler, 1973) probably represent other xerothermal relicts.

The divergence between California and Arizona segments of *C. magna* probably dates from the Quaternary development of a Mediterranean climatic regime in California, but there may have been limited gene flow between these areas during the Pleistocene fluctuations. Pleistocene connections with both Arizona and California could account for the phenetically central position of the Providence Mountain beetles, which show relatively high similarities to individuals from these areas as well as the Great Basin. There is some indication that the northern and southern cismontane California populations of *C. magna* may have been periodically separated during the Pleistocene. Although continuous populations now exist from northern Baja California to at least the San Francisco Bay region and probably to the northern end of the Sacramento Valley, there is a suggestion of a discordance just north of the Transverse Ranges. Single character analysis of localities from this region reveals gradual changes along a north-south gradient (figs. 12–37) but overall phenetic similarities tend to be slightly higher among the northern and southern localities (figs. 44–45). Temporary isolation or greatly reduced gene flow between the northern and southern populations may have occurred during the Xerothermic Period, and even today the distribution of *Coelocnemis magna* just north of the Transverse Ranges is rather fragmented, corresponding to the patchy occurrence of Woodland habitats. Thus, colonies which appear to be isolated occur in the Tehachapi Range (Split Mountain), west of McKittrick and along the Cuyama River Valley. If separations of the northern and southern segments of *magna* did take place they were too brief to allow development of effective isolating mechanisms and a broad zone of intermediacy now extends from the Transverse Ranges to nearly the San Francisco Bay region.

The populations designated as *C. sulcata* consist of numerous isolates in the mountains of the Basin and Range Province. The Great Basin populations occupy

---

[4] The distribution of *Gerrhonotus* is almost precisely the same as that of *C. magna*. Besides the isolate in the Panamint Range there are disjunct populations of the lizards on Cedros Island and in extreme southern Baja California, and both are considered distinct species. Woodland areas of both Arizona and California are occupied by separate species, and the Sierra Madre Oriental support yet another species which corresponds geographically to the tenebrionid genus *Oenopion*.

semiarid pinyon-juniper woodland habitats and occur sympatrically with *C. punctata* in most of their range. The Arizona populations occur in much more mesic pine–oak-woodland areas in Arizona, where they are closely adjacent to populations of *C. magna* in several cases, but apparently not in actual sympatry.[5]

The morphological relationships between *magna* and *sulcata* are confusing. Although the Great Basin and montane Arizona populations of *sulcata* share a common pattern of elytral sculpturing they show minor differences in several characters which measure punctation and body proportions. On the basis of these other characters the Great Basin populations of *sulcata* are similar to Owens Valley and northern California representatives of *C. magna*, while Arizona *sulcata* is more similar to *magna* populations from Arizona. The very weakly sulcate individuals from the Providence Mountains, California, have relatively high similarities to specimens from all these areas. This evidence can be interpreted in more than one way. One possibility is that *C. sulcata* is actually polyphyletic, the Great Basin populations deriving from California populations of *magna* and the Arizona populations deriving from Arizona *magna*. Alternatively, if monophyletic, *sulcata* could have been derived from beetles in either California or Arizona, with subsequent morphological divergence. Since the paleoclimatic evidence does not favor either of these hypotheses, all the sulcate populations are considered to represent a single polytypic species.

Two anomalous races of the *Magna* group remain to be discussed. First, in the Mojave River Valley near Victorville exists an apparently isolated population of beetles which are phenetically most similar to populations from San Luis Obispo County. Since the Victorville population is known from very few specimens this unexpected similarity may be due to sampling error, and it is possible that the Victorville area has been colonized from the adjacent San Bernardino Mountains in historic time. Second, a distinct race of *magna* occupies restricted mesic areas along canyon bottoms on Santa Catalina Island. These beetles are very similar in body proportions to specimens from the adjacent mainland, but differ markedly in cuticular sculpturing. It is not clear whether Santa Catalina was ever connected to the mainland, although Pliocene connections to Santa Cruz Island have been suggested (Corey, 1954; Valentine and Lipps, 1967). It is not possible to specify the time of arrival of *Coelocnemis* on the island, but the close proximity of Santa Catalina to the mainland suggests that colonization may have occurred at any time since the Pliocene. Interestingly, *Coelocnemis* is absent from Santa Cruz Island, which supports a more mesic flora than the other California islands and boasts a considerable fauna of Californian tenebrionids, including *Cibdelis, Nyctoporis, Coniontis, Eulabis, Phloeodes, Noseris, Cratidus,* and *Eleodes* (sub-

---

[5] In the Santa Catalina Mountains *C. magna* is common on the south slope above about 1,500 m, while *sulcata* has been taken from the north slope at similar elevations. Whittaker and Niering (1968) compare the vegetation of the north and south faces of the Santa Catalinas, finding distinct vegetational as well as edaphic differences. The south slope shows floristic intergradation at middle and lower elevations with Sonoran desert elements and consists primarily of granitic gneiss; the north slope exhibits greater similarity with the Chihuahuan flora and has a varied substrate of granite, limestone, diorite, shale, quartzite, and other rocks. These differences suggest that *magna* and *sulcata* have somewhat different ecological tolerances; in general *C. sulcata* seems to be adapted to slightly more arid situations than *magna*.

genera *Eleodes* and *Blapylis*). Several of these genera are represented by species or races endemic to Santa Cruz, Santa Catalina, and other channel islands, suggesting the possibility that *Coelocnemis* may be a relatively recent arrival on Santa Catalina, and has never occupied the other islands.

THE CALIFORNICA SPECIES GROUP

The paleoclimatic changes which influenced the distribution of the Magna species group would have had an inverse effect on the Californica group. The large area of western conifer vegetation in the Pacific Northwest presently supports morphologically homogeneous populations of *Coelocnemis californica*, probably indicating a postglacial invasion of this region. In the southern extremes of its distribution, however, a number of distinctive races and species are present, reflecting the late Cenozoic and Pleistocene fluctuations in temperature and rainfall as well as the profound geological changes there. The San Francisco Bay region and montane Southern California races are differentiated by rather minor morphological peculiarities, probably caused by periodic isolation during the Pleistocene. At the present time the San Francisco Bay region populations occur in isolated mesic areas at least as far south as the Santa Cruz Mountains (there is a single, unverified record from the northern Santa Lucia Range). South of San Francisco Bay these populations are largely allopatric, but from Marin County northward the beetles are continuously distributed and intergrade with the north coastal populations. It is apparent that during the various glacial maxima the presently interrupted distribution must have been more continuous and probably extended south into the Santa Lucia Range. Axelrod (1967) cities evidence that *Pseudotsuga menzesii* (Mirb) Franco, *Sequoia sempervirens* D. Don., *Cupressus goveniana* Gordon, and *Garrya elliptica* Douglas, now more northerly in distribution, were present on Santa Cruz Island and on the mainland near Carpinteria from about 14,000 to 38,000 B.P. This would indicate a southerly shift in distributions of about 200 miles, which would probably have brought *C. californica* and *C. lucia* into sympatry, assuming that the latter was not displaced to the south. The taxonomic implications of these changes in range are discussed below. It seems clear that the San Francisco Bay region populations have never been isolated from the beetles to the north for extensive periods of time or by very great distances and that San Francisco Bay and the Sacramento delta have not been significant barriers to dispersal. One may also speculate that the distinctive phenotype of the Bay region populations originated during the Pliocene submergence of the north coastal region in the vicinity of the Russian River, which isolated the Marin County mountains and low ranges of mountains east of San Francisco Bay. The middle Pliocene, however, was also the time of greatest northerly expansion of Madro-Tertiary elements in California, and it is problematical whether the climate would have been suitable for *C. californica*.

The extension of *C. californica* into the southern California mountains probably took place no earlier than late Pliocene when the first invasion of Arcto-Tertiary plants arrived (Axelrod, 1958*b*). The southern California populations are very similar to those of the southern Sierra Nevada, differing primarily in the shape of the prosternal process. Their distribution in the mountains of southern California is typical of Arcto-Tertiary relicts, with isolates in the higher ranges. Very small but typical populations in moist canyon bottoms in the Sierra Madre of Santa Barbara County record the more mesic conditions of the western Transverse Ranges in the recent past. The present isolation of the southern California beetles from those in the Sierra Nevada has undoubtedly persisted since the Xerothermic Period (8000–3000 B.P.). Between this time and the late Pliocene there were probably repeated faunal connections corresponding to the Pleistocene glaciations and accounting for the phenetic similarity of the existing Sierran and southern Californian populations.

While the present distribution of *C. californica* is understandable in terms of Pleistocene geological events and climatic fluctuations, the occurrence of the distinctive *C. lucia* in the Santa Lucia Range demands another explanation. These beetles are phenetically distinct from both the San Francisco Bay region and southern California populations; their highest similarities are to specimens from north coastal California, from which they differ mainly in their much smaller size. Geologically the Santa Lucia Range has been relatively stable, escaping the Miocene and Pliocene submergences which affected central California to both the north and south. During the Pliocene the Santa Lucia Range was periodically isolated as an archipelago, but since the middle Pliocene connections have existed with the continental land mass to the north. Isolation from the southern California land mass was effected by a broad strait connecting the San Joaquin embayment with the Pacific Ocean. This barrier persisted until mid-Pleistocene time. Peabody and Savage (1958) believe that the peninsular condition of the outer Coast Ranges during much of the Pliocene and Pleistocene explains some of the distributional phenomena of the California herpetofauna, particularly the high incidence of endemism in southern California and the low endemism in the Santa Lucia Range. The presence of a marine barrier just north of the Transverse Ranges reinforces the hypothesis that *C. magna* populations in this region have only recently come into contact. By similar reasoning, *C. lucia* must have entered from the north, but its relatively great phenetic divergence from *C. californica* suggests that this occurred well before the Pleistocene glaciations responsible for the range extensions into southern California. Possibly *C. lucia* became established in its present location by the early Pliocene and was isolated by the mid-Pliocene submergence of the Salinas Valley and Monterey area. By the late Pliocene when cooling temperatures and emergence of the Salinas Valley permitted contact with beetles from the north, *C. lucia* had apparently evolved the features which separate it from *C. californica*. Although *lucia* and *californica* are allopatric at present (possibly sympatric in the northern Santa Lucias) they were probably sympatric during the glacial maxima of the Pleistocene. It is primarily because of this suspected sympatry that *C. lucia* is treated as a separate species rather than a race of *C. californica*. Peabody and Savage believe that the Santa Lucia Range was floristically Madro-Tertiary in character until the Pleistocene, but the presence of *C. lucia* indicates that relatively mesic areas have probably persisted locally since the Pliocene. The low incidence of endemism in the herpetofauna also contrasts with the situation in certain other groups of organisms. Stebbins and Major (1965) show that the central Coast Ranges in California are relatively rich in endemic species of plants and they include the Santa Lucia Range as one of a number of local areas of concentrated endemicity. The central Coast Ranges are also relatively rich in endemic microlepidoptera (J. A. Powell, personal communication), and the tenebrionid beetle *Cibdelis laevigata* Casey is endemic to the Santa Lucias and the western Transverse Ranges in Santa Barbara County. The Santa Barbara County records may represent a Pleistocene adjustment of distribution similar to those postulated by Peabody and Savage for a number of amphibians and reptiles. Conversely, neither *C. californica* nor *C. lucia* seem to have crossed the gap between the Santa Lucias and the Transverse Ranges since the mid-Pleistocene subsidence of the southern strait. Except for the discordant Santa Lucia Range beetles, the populations of *C. californica* would form a nearly complete circular cline around the central valley, much as in *Ensatina eschscholtzii* (see Stebbins, 1949, 1957 for details).

In contrast to the complex situation in California, the populations of *C. californica* extending into the Wasatch and Uinta Mountains of Utah are similar in all characters to the undifferentiated populations in the Pacific Northwest. This reinforces the idea that the Pacific Northwest was invaded in postglacial times by populations in northern California. The absence of *C. californica* from the Colorado Rockies provides further evidence of the recency of colonization in this area. Findley and Anderson (1956) compare the recent distributions of a number of mammalian species in this area and show that the Green River and Wyoming Basins act as effective barriers to many of them. They further conclude that both these barriers are probably Pleistocene in origin but that the Green River Canyon is more recent than the Wyoming Basin. If these ideas are correct, *C. californica* probably did not invade the Uinta and Wasatch mountains until at least mid-Pleistocene times, explaining its absence from the Colorado Rockies and its lack of phenotypic divergence.

The Great Basin and Colorado Plateau populations of the Californica species group constitute the species *C. punctata*. The Great Basin populations are rather uniform in most characters but vary significantly in size, sculpturing, and stoutness of the body. At the western edge of its range *C. punctata* assumes a phenotype suggesting Sierra Nevadan populations of *C. californica*. A few specimens from Craters of the Moon National Monument on the Snake River plain in Idaho also show intermediacy, but in Utah where *californica* and *punctata* occur in geographic contiguity if not in actual sympatry, no intermediacy is known. In Modoc County, California, *punctata* occurs on the floor of Surprise Valley, less than 10 km from (but 900 m lower than) Warner Mountains localities for *californica*. Near or actual sympatry may occur in other areas as well, such as Reno, Nevada. It is not possible to deduce the probable time of divergence of *punctata* from *californica* on the basis of paleoclimatic information. However, available evidence indicates that pinyon-juniper woodland was more extensive during at least the last glacial maximum, probably interconnecting many of the presently isolated stands. According to the interpretation of Wells and Berger (1967) xeric plants apparently dominated the lowlands at this time and montane species such as ponderosa pine and white fir were nearly as limited as today, judging from their present uneven distribution in the basin ranges. According to these ideas the insular character of woodland plant associations persisted in the Great Basin throughout the Pleistocene, and *C. punctata* could conceivably have invaded the basin ranges during this time. Perhaps it is more likely that *punctata* was already established in the Madro-Tertiary woodlands or in ecotonal areas by mid-Pliocene time. In either case the repeated fluctuations during the Pleistocene seem responsible for the presently fragmented distribution. The conspicuous gap in northwestern Nevada (fig. 51) corresponds to the location of pluvial Lake Lahonton.

The Colorado Plateau race of *punctata* differs from the Great Basin form in color and elytral sculpturing, but specimens from the Henry Mountains and other localities in south central Utah are intermediate in both these characters. A continuous cline may have existed in this area during more mesic times, with free gene flow between the two races. Unfortunately collections from this region are too scarce to allow a reliable assessment of present phenetic relationships.

The species *C. rugulosa,* endemic to the Modoc Plateau and adjacent areas of extensive recent vulcanism, remains to be discussed. This species is sympatric with *C. californica* throughout most of its range, but tends to occur in lower, more arid localities dominated by juniper and subclimax yellow pine. The only record of *rugulosa* west of the Cascade Range is from the closely adjacent Siskiyou Mountains of southern Oregon. Apparently this species became adapted to relatively arid, intemperate conditions during some period of isolation from *californica,* but it is not possible to deduce from the available evidence what geological or climatic barriers might have contributed to its evolution.

Summary of Biogeographic Analysis

It should not be supposed that the above documentation is sufficient to reconstruct a reliable phylogeny of *Coelocnemis,* and the classification adopted here is not based on phylogenetic evidence (see discussion, pp. 81–84). However, it seems clear that the present distribution and major patterns of variation of these beetles conform remarkably well with the proposed limits of Axelrod's geofloras.

The remotely isolated populations on Cedros Island and the cape region of Baja California, in particular, seem to support the theory that a woodland vegetation not greatly different from that now present in southern California was much more widespread in pre-Pleistocene times. The presence of phenetically similar forms of *C. magna* in the pine–oak woodland of both California and Arizona also parallels the distributions of a number of extant and fossil plants and seems to support Axelrod's reconstructions. The evidence from other Tenebrionidae, however, is less conclusive. The California oak-woodland and chaparral formations, supposed modern derivatives of the Madro-Tertiary Geoflora, support a number of endemic Tenebrionidae. The genera *Nyctoporis, Phloeodes, Cibdelis, Eulabis,* and *Cratidus* are each represented by several species or races which occur in the same habitat as *C. magna,* and all may frequently be taken from the same snag or fallen tree. None of these genera occur in Arizona, where the tenebrionid associates of *Coelocnemis* are largely species of *Eleodes,* and only *Cratidus* is known from Cedros Island or southern Baja California. Large genera such as *Eleodes* and *Coniontis,* which are too complex to consider here, may show close Arizona–California correspondences when their relationships are unraveled, but present evidence does not suggest a close relationship between tenebrionid communities from these areas.

Most of the disjunction in the distributional patterns of *Coelocnemis* species suggests minor extensions and contractions of range stimulated by Pleistocene climatic fluctuations. However, it seems likely that the main patterns of morphological variation were established during earlier periods and that only minor divergences occurred as a result of Quaternary events. The Great Basin and desert mountain range distribution of *Coelocnemis* supports a moderate Pleistocene displacement of vegetational zonation in accord with the interpretations of Wells and Berger (1967), Johnson (1968), and Howden (1966).

SYSTEMATICS
GENERIC RELATIONSHIPS

The taxonomic conclusions presented below are derived largely from the phenetic and biographic evidence already discussed, but several additional characters are taxonomically important, particularly in determining generic relationships.

Genitalic differences were not included in the previous analyses because of the extensive amount of dissection and preparation that would have been required. However, examination of more than 300 specimens of *Coelocnemis* as well as selected individuals of *Cibdelis, Scotobaenus, Iphthimus, Alobates, Oenopion,* and *Rhinandrus* revealed both intergeneric and interspecific differences. In *Scotobaenus* the ovipositor is highly sclerotized and bladelike, but in the other genera it is largely membranous, similar to that of *Tenebrio*. The female genitalia of *Coelocnemis* are nearly identical with those of *Iphthimus* and *Oenopion;* morphological differences among the remaining genera are minor with the exception of *Cibdelis,* in which the ovipositor and spiculum are extremely elongate. In general these structures are of limited utility in clarifying relationships and are not discussed further here.

The male genitalia are much more diverse, but *Coelocnemis, Iphthimus,* and *Oenopion* again share a high degree of similarity. The aedeagus is of moderate size, heavily sclerotized except ventrally, and strongly curved in the dorsoventral plane. The penis is well sclerotized, short and sessile, and hinged to the aedeagus at the juncture of the basal piece and paramere, from which it extends like the blade of a pocket knife. The remainder of the genera examined differ chiefly in the shape of the aedeagus and the relative proportions of the basal piece and paramere. *Scotobaenus* is again divergent, with a series of lateral teeth on the paramere and a strong apical hook on the penis. In *Rhinandrus* the aedeagus is very small relative to body size and in *Alobates* the paramere is much smaller relative to the basal piece than in the other genera. The male structures in *Cibdelis* are similar to those of *Coelocnemis*.

Within *Coelocnemis* differences in structural details of the aedeagus and penis further substantiate the two groups of species alluded to throughout the text. In the Magna group (fig. 52) the aedeagus and penis are relatively elongate and slender, the basal piece is sharply curved proximally and the posterior processes of the paramere are fused. There are only subtle differences among the species of this group. The Californica species group (fig. 53) is characterized by a shorter, thicker aedeagus and penis. The basal piece is more gently curved and the pos-

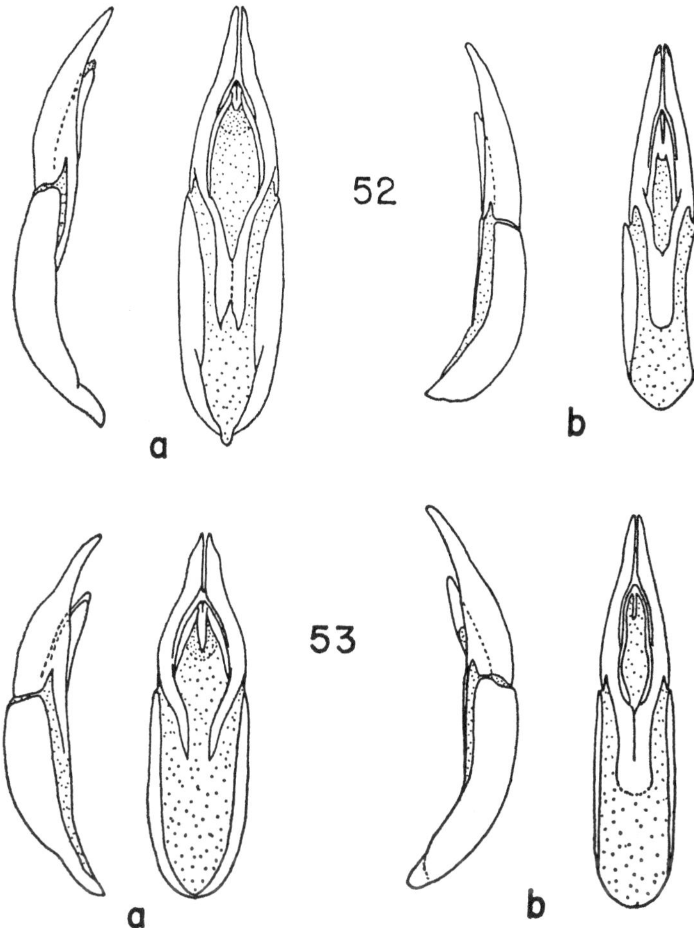

Figs. 52–53. Male genitalia of *Coelocnemis*, dorsal (left) and lateral (right) aspects. Fig. 52. (a) *C. magna* LeConte; (b) *C. sulcata* Casey. Fig 53 (a) *C. californica* Mannerheim; (b) *C. punctata* LeConte.

terior processes of the paramere are separate. *C. punctata*, including the Colorado Plateau populations (*C. angusta* Casey), is an exception in having the processes of the paramere broadly fused. This fact supports the inclusion of all the Great Basin and Colorado Plateau specimens in a single polytypic species in the present classification. The remaining Californica group species show only minor variation in genitalic structure.

From this limited survey it is apparent that external genitalia, especially of the males, offer important characters for the higher classification of the Tenebrioninae, as indicated by Koch (1960). Among the taxa considered here, *Coelocnemis* is very similar to *Iphthimus* and *Oenopion*, but a number of genera (*Coelometopus, Astathmetus, Hipalmus*) which may be critical to this study have not been examined.

Coelocnemis is distinguished from other North American genera of Tenebrionini and Coelometopini except *Iphthimus* by the large, platelike mentum. In *Cibdelis* the mentum is moderate in size. In the other genera examined, including *Oenopion*, it is relatively small with distinct lateral lobes. This is one of the chief reasons that *Oenopion* is considered a separate genus in the present work. The unusual structure of the mentum was noticed by Lacordaire, who originally classified *Coelocnemis, Cibdelis,* and *Polypleurus* as Scaurini, but later (1859) erected a new tribe, the Coelometopini, chiefly on the basis of the apterous body and tarsal vestiture. Besides the three genera just mentioned, LaCordaire's "Coelometopides" originally included *Centronopus* and *Coelometopus,* the latter Palearctic in distribution and very similar to *Iphthimus.* Papp (1961) places nine Nearctic genera in the Coelometopini, including *Coelocnemis, Cibdelis, Oenopion, Centronopus,* and *Scotobaenus* but not *Iphthimus.* Neither of these groupings is substantiated by the phenetic evidence presented here. The apterous body is a specialization that has occurred repeatedly in the Tenebrionidae; any classification utilizing this character will be polyphyletic. This is particularly true in the Tenebrionini and related tribes, in which both winged and wingless genera are commonly present. The structural specializations of the genitalia make it difficult to fit *Scotobaenus* and the closely related genus *Centronopus* into existing classifications. The Tenebrionini and Coelometopini might conveniently be classified according to the free or sessile condition of the penis, but many more genera in these and other tribes must be examined before the value of this character can be verified. Tentatively, *Coelocnemis* may be assigned to the Coelometopini.

Characteristic setal patterns on the legs are common in genera of Tenebrionini and Coelometopini, frequently as secondary sexual characters. In *Coelocnemis* the pattern consists of paired parallel lines of fine, yellow pubescence on the apical half of the tibiae, a configuration unique to known genera except *Oenopion.* The latter differs in having an elongate patch of pubescence on the ventral surface of the femora in males. In general the presence of setal patterning on the legs of Tenebrionidae and closely related families is positively correlated with a preference for a substrate of dead wood. The genera *Nyctoporis, Phloeodes,* and *Zopherus,* which all occur on and around tree snags, are peculiar among the genera of the Tentyriidae (= Tentyriinae of Authors; see Doyen, 1972) in having the tarsi covered with silken pubescence. *Zopherus* also bears parallel lines of pubescence on the tibiae and femora. Although patterns of pubescence are evidently an important adaptive feature in tenebrionids preferring a ligneous substrate, the biological significance to the individual is not known.

Substrate differences are also correlated with larval morphology. Larvae which burrow in decomposing wood frequently have long, recurved urogomphi, as in *Coelocnemis, Iphthimus, Centronopus, Scotobaenus, Merinus,* and *Haplandrus.* Larvae which inhabit friable substrates usually lack urogomphi, as in many species of *Eleodes,* or show short posteriorly directed spines, as in *Tenebrio, Tribolium, Nyctoporis, Coelus, Coniontis,* and some species of *Eleodes.* Hayashi (1966, 1968) provides many more examples of the correlation between large urogomphi and ligneous larval habitat. The larvae of the tribes Tenebrionini and Coelometopini are frequently characterized at the generic level by the structure of their urogomphi as well as other characters. However, the distribution of larval characters among genera presents a mosaic pattern similar to the occurrence of components in the defensive secretions of adults, and intergeneric relationships are confusing and probably involve convergence. The hemicircle of complexly wrinkled spines of *Coelocnemis* is unique among known tenebrionid larvae, but in most larval characters *Coelocnemis* is similar to *Iphthimus* (see St. George, 1924; pl. 3–4). The elaborate terminal armature of *Coelocnemis* is approached in the larvae of the genus *Strongylium* (see Hayashi, 1966), but there are differences in mouthpart structure in the adults. The larvae of *Oenopion* and *Cibdelis*, probably important to any phylogenetic speculations based on immatures, are unknown.

Larvae and pupae of the various species of *Coelocnemis* are extremely uniform morphologically. Differences in the terminal armature and pigmentation patterns vary as much between individuals of a single population as between species. Consequently, immatures have not been used in constructing the classification presented below.

Integration of Statistical Methods

The complexities of morphological variation discussed at length in the preceding text should make it clear that any classification of *Coelocnemis* will be arbitrary in some respects. This is partly because each method of analysis presents a different combination of phenetic relationships, but also because of the gradual nature of the observed variation and the numerous discordant characters.

In some instances the differences between methods of analysis are not remarkable (e.g., correlation coefficient and taxonomic distance phenograms), but major disagreements are frequent (e.g., among phenograms, taxometric maps, and "phenomaps"). Additionally, the trends indicated by STP analysis must be accommodated. Sokal (1969) believes that the problem of integrating the results obtained from different coefficients of similarity will be alleviated in the near future as mathematical statisticians develop more robust, monotonic coefficients. He also postulates that it may be possible to tailor the clustering procedure to the phenetic structure of the pertinent taxonomic group. Thus, it would be inappropriate to cluster the members of a continuous cline by average linkage methods. For *Coelocnemis* this would necessitate using *different clustering methods for various portions of the genus*. The clinally related populations of *C. magna* in cismontane California would be best represented by single-linkage techniques or nonhierarchical methods such as taxometric mapping, whereas the relationships of phenetically disjunct populations might be suitably displayed by average-linkage phenograms. It is not clear which methods would be appropriate for certain OTUs (SYAV, RETC, CAPE) which cluster erratically with different coefficients and clustering techniques. At any rate it is apparent that the stability promised by numerical taxonomy has not yet been attained, a conclusion shared by other workers (Michener, 1970; Eades, 1970; Hull, 1970). Even if monotonic similarity coefficients are developed the subjective judgment of the taxonomist will remain responsible for evaluating the underlying phenetic structure of a taxonomic group and matching it with the most appropriate numerical methods.

It is desirable now to review in detail the discordances apparent in the various approaches utilized in this study. Reexamination of the STP diagrams will show that single-character variation is predominantly gradual, with significant differences usually present only between populations at opposite ends of the ranges of means. For example, male pronotal length is significantly different between most populations of *C. magna* from Arizona and cismontane California, but RCHR (Arizona) belongs to the same subset as RCAL, OBAJ, OLAN, and MCCO (California). This overlap in values is repeated for every character except color and elytral sulcation, and contrasts sharply with the discontinuities indicated by multivariate analysis. This difference is partly due to actual phenetic gaps which appear only when all the characters are analyzed simultaneously, but also reflects the fact that multivariate techniques cluster the OTUs only on the basis of the mean character values without regard to statistical significance. Univariate analysis reduces the apparent size of the gaps because it also takes into account the intrapopulation variation of each character. Realistic appraisal of the relationships derived from multivariate methods requires review of the STP diagrams to determine the degree to which each character contributes to the overall phenetic differences and how important these differences actually are.

One method of accomplishing this might be to weight each character in proportion to the average number of significantly different subsets included in its overall range of variation. For example, *pronotal length* in females of the Magna species group (fig. 12) has eight subsets, containing an average of 12.25 samples each. There are 32 samples in the Magna group so that for pronotal length the average number of subsets is 32/12.25, giving a weighting factor of 2.6. A character with less variation, such as *prosternal rugosity* (fig. 33), would receive a weight of only 1.3. A similar procedure, basing weights on the number of standard deviations within the range of each character appears as an option in Carmichael and Sneath's (1969) taxometric mapping method.

Comparison of the results of the various multivariate techniques applied to *Coelocnemis* shows that each reveals different aspects of the phenetic relationships within the genus. As mentioned above the members of the continuous cline in cismontane California are best represented by single-linkage procedures (phenomaps or single-linkage phenograms). This is a very clear cut example, however, and it is not clear which method most accurately portrays the relationships of the phenetically variable California populations of *magna* or the *sulcata* populations. In the phenograms these OTUs show confusing relationships among themselves as well as with cismontane California and Arizona populations of *magna*. In the phenomaps the *sulcata* samples are strongly interlinked, but SPRO shows high similarities to transmontane California samples in both sexes. In the taxometric maps both groups of populations appear mostly as isolated OTUs, with *sulcata* populations interconnected predominantly among themselves. The relatively low similarities among the *sulcata* populations may be due in part to the great geographic distance between sample localities, especially in the Great Basin. However, the distances separating the Arizona localities for *magna* are also large, while the similarities are high. Probably the phenomaps, which show only the two highest similarities, most accurately represent the interrelationships of the Great Basin and desert mountain range populations of the Magna group.

In the Californica group the San Francisco Bay region and southern California localities cluster separately in the phenograms and the latter in the phenomaps as well. The taxometric maps separate the southern California samples (in a single cycle of clustering) only in the males and the bay region OTUs are not separated in either sex. While these populations are somewhat distinct, the phenetic gaps separating them from Sierran and north coastal California populations, respectively, are magnified by hierarchical analysis. They are best clustered by the nonhierarchical procedure employed in taxometric mapping. The same considerations apply to the Great Basin populations (*C. punctata*), but in this case the phenetic gaps are larger and involve more characters.

Four populations (CLUC, RUGU, CEDR, CAPE), are isolated by all methods of multivariate analysis. These populations also show significant differences from most other OTUs in several of the STP diagrams, typically appearing at one extreme of the observed variation. They are considered as distinct species in the classification presented below.

PHENOTYPIC DIVERGENCE AND TAXONOMIC RANK

Aside from the continuously distributed beetles in cismontane California, the Magna species group consists of allopatric, morphologically distinct populations. This is the classical situation which has stimulated the recognition of subspecies in many groups of organisms, and designation of subspecies would accommodate much of the variation in *C. magna,* although the resulting classification would be extremely complex. Such a categorization would also impose a deceptively static complexion on the intricate phenetic relationships within the genus. In both species groups the degree of phenotypic divergence ranges from a few barely discernible differences between individuals at opposite extremes of the variation to populations separated by several characters with little overlap. There is no consistent way to fit this spectrum of relationships into a subspecies classification. For example, most individuals of *C. magna* from Arizona are separable from those from California and they differ in the chronology of their life cycles as well. These populations seem to merit taxonomic recognition, but if the numerous populations isolated in the intervening desert mountain ranges are taken into account only arbitrary decisions can be made. The Sierra San Pedro Martir populations could be included with either the Arizona or California populations or given independent status, depending on which characters were emphasized. The populations in the Owens Valley and Little San Bernardino Mountains pose a similar problem, differing in a few characters from both the Arizona and California phenotypes, but with high similarities to both. Additionally, several of the geographically continuous populations are themselves variable. In *C. magna* the degree of divergence between beetles from the northern and southern extremes of the range in cismontane California is as great as that between allopatric populations, but the continuous nature of the variation in the former does not permit taxonomic recognition. It seems likely that most of the presently allopatric populations were isolated by recent geologic and climatic events. Temporary isolation of this sort has probably accompanied all of the major events of the climatically and geologically complex Pleistocene period. It is likely that the contemporary allopatry and accompanying minor phenotypic diversity of *Coelocnemis* represent relatively recent adjustments to climatic changes rather than major evolutionary trends of long duration.

Examination of the Californica species group reveals another difficulty. While the allopatric distribution of the Magna group could be adapted to a classification into subspecies, nearly all the variation in the Californica group is gradual and geographically continuous. If allopatry is used as the primary criterion for subspecies recognition, *C. magna* would be split into numerous slightly differentiated entities, while *californica* would be divided into one homogenous subspecies in the southern California mountains and another extremely variable subspecies distributed throughout the rest of the range of the species. Recent detailed analyses of diverse groups of animals (some examples are: James, 1970, birds; Johnston, 1969, birds; Johnston and Selander, 1964, birds; Jolicoeur, 1959, mammals; Lidicker, 1960, mammals; Linsley and Chemsak, 1961, insects; Power, 1970, birds; Rising, 1970, birds) indicate that intricate patterns of geographic variation involving gradual changes and frequent character discordance are of common occurrence. The taxonomic treatment in the studies listed above varies, but it seems clear to me that, at least in *Coelocnemis*, the observed variation cannot be realistically reflected in a classification of subspecies.

Most studies of geographic variation include only single species or a portion of one species. This may explain the preoccupation with the difficulties associated with subspecies. The present work, involving all the available material of an entire genus, suggests that the same problems are common also to the rank of species. As set forth by Ehrlich and Raven (1969) and elaborated by Sokal and Crovello (1970), the biological species concept is almost never applicable to natural populations, which are classified phenetically in practice. The Providence Mountains population of *Coelocnemis* provides an illuminating example. These beetles are morphologically intermediate between *sulcata* and *magna*, making it impossible to place them objectively in either species without detailed biochemical or genetic evidence. Even with this additional information, the evolutionary relationships of these populations would probably be too complex to be represented accurately in a rigid, hierarchial system of nomenclature. Their sulcate elytra make it convenient and practical to include the Providence Mountain beetles in *C. sulcata*, a decision made purely on the basis of phenetic evidence.

Another example is provided by *C. californica* and *C. punctata*. Populations in the Owens Valley and a few other areas are phenotypically intermediate between these two "species," but in general they are easily differentiated. Although they are almost entirely allopatric, suggesting that only a single species is present, *punctata* and *californica* have each diverged into several geographic races, making it more convenient to consider them as separate species. As in the former example this decision is made on phenetic grounds.

In presenting these examples I have not intended to criticize the biological species as a useful *evolutionary concept*. Nevertheless I agree with Ehrlich and Raven (1969) that the genetic relationships among populations of closely related organisms are apt to be so complex that any classification will be only an approximation of the actual situation. Taxonomic classifications should only attempt to reflect the major phenetic trends in groups of organisms, unless explicit evidence regarding their evolutionary history is available.

TAXONOMIC DECISIONS

The classification presented below is based on the phenetic relationships apparent from the various methods of quantitative analyses, and also takes into account the historical events discussed in the section on biogeography. The species recognized here are mostly polytypic aggregations of local geographic races. Because of the objections to subspecies mentioned in the previous section, geographic variation is treated by describing phenotypically distinct series of populations, but these are accorded no nomenclatural status. In several instances gradually varying populations are broken into races if regional differences allow them to be characterized. In these cases the character intergradation is briefly summarized. Complete accounts of the variation may be obtained by reviewing the sections on quantitative analysis.

The Magna Species Group

This group consists of three species, *C. magna* LeConte, *C. sulcata* Casey, and *C. slevini* Blaisdell. The beetles from extreme southern Baja California clearly represent a distinct species on phenetic grounds, but it is not described here because of insufficient material.

C. magna includes all the nonsulcate specimens of this species group, with the exception of those from Cedros Island. The northern and southern California phenotypes were considered separate species by LeConte and later workers, but there is no justification for recognizing more than a single species, since every character in which they differ is subject to clinal variation, with a broad zone of intermediacy from Monterey County south to the Transverse Ranges. The Arizona populations of *magna* have also been described as a separate species, and with greater justification, since a distinct phenetic gap separates them from the California populations. However, a substantial minority of specimens is indistinguishable from material from central cismontane California. Furthermore, the Owens Valley populations are clearly intermediate between Arizona and California specimens in prothoracic dimensions and punctation. For these reasons all are considered as a single species. The populations isolated in the Little San Bernardino Mountains and on Santa Catalina Island are phenotypically distinct, but only in a few characters, and do not warrant recognition as species. They were probably derived from geographically adjacent populations in relatively recent geological time. The Sierra San Pedro Martir specimens show ambiguous similarities to both Arizona and southern California populations. Paleoclimatic evidence suggests that they were probably isolated from southern California populations during the Pleistocene.

C. sulcata Casey contains all the sulcate specimens of the Magna species group. There are minor differences in body proportions between Great Basin and Arizona races, and some suggestion that *sulcata* may be polyphyletic. The present classification is based primarily on elytral sulcation, the only character in which all the sulcate individuals differ from *magna*. The Providence Mountains population is included in *sulcata* for this reason. In several other characters it is most similar to Owens Valley specimens of *magna*.

C. slevini Blaisdell is endemic to Cedros Island and phenetically distinct from all other populations. In most cuticular sculpturing characters it falls at one extreme of the observed variation and differs in the structure of the mentum and prosternal process. Its phenotypic distinctness and the paleoclimatic history of Baja California indicate that it has been isolated for a long period of time, justifying its recognition as a species.

The Californica Species Group

The populations of this species group are considered to represent four species, *C. californica* Mannerheim, *C. punctata* LeConte, *C. rugulosa* n. sp., and *C. lucia* n. sp.

C. punctata LeConte contains the Colorado Plateau and Great Basin specimens, including material from the eastern escarpment of the Sierra Nevada, the Mono Plateau, and a few intermediate individuals from the Snake River Plain in Idaho (Craters of the Moon National Monument). The Colorado Plateau specimens are aberrant in color and elytral sculpturing, but similar to typical *punctata* in other features, including genitalia. Their differentiating characters probably represent an adaptation to the peculiar edaphic conditions of the Colorado Plateau. All methods of analysis except taxometric mapping (females) show a distinct phenetic gap between *punctata* and *californica*. Along with its differentiation into distinct geographic races this warrants its recognition as a species, although it is almost completely allopatric with *californica* and seems to intergrade with it in the areas listed above.

C. rugulosa and *C. lucia* are known from phenotypically distinct populations from the Modoc Plateau and the Santa Lucia Mountains, respectively. *C. rugulosa* is presently sympatric with *californica* and paleoclimatic evidence suggests that *lucia* and *californica* were sympatric during the Pleistocene glacial maxima. These populations would be considered races of *californica* on phenetic evidence alone.

C. californica includes all the remaining specimens of this species group. This species is distributed over a very large area and has differentiated into several geographic races which have previously been considered distinct species. Since the races intergrade in all differentiating characters there is no justification for retaining them as species. Zoogeographic evidence indicates that they have probably been isolated only since the Pleistocene glaciations.

Genus *Coelocnemis* Mannerheim

Coelocnemis Mannerheim, 1843, Bull. Soc. Imp. Nat. Moscou, 16:280; Mannerheim, 1844, Mag. Zool., 6 (Ser. 2), Part 133, p. 1; LeConte, 1851, Ann. Lyceum Nat. Hist. N.Y., 5:150; LeConte, 1854, Proc. Acad. Nat. Sci. Phil., 7 (185)5:225; Lacordaire, 1859, Histoire naturelle des Insects. Genera des Coléoptères, 5:363; Horn, 1871, Trans. Amer. Phil. Soc., 14:335; Gissler, 1879, Bull. Brooklyn Entomol. Soc., 2:7 (biology); LeConte and Horn, 1883, Smithsonian Misc. Coll., 26:377; Casey, 1895, Ann. N.Y. Acad. Sci., 8:615; Linnell, 1901, Proc. Entomol. Soc. Wash., 4:181; Casey, 1924, Memoirs on the Coleoptera, 11:314; Blaisdell, 1925, Proc. Calif. Acad. Sci., 14:337; St. George, 1924, Proc. U.S. Natl. Mus., 65:4 (larva); Blaisdell, 1928, Pan-Pac. Entomol., 15:163; Blaisdell, 1943, Proc. Calif. Acad. Sci., 24:273.

Males.—Moderately robust, apterous, uniformly black or infrequently reddish black.

Head dorsally flattened, approximately hexagonal in dorsal aspect, widest at antennal insertions, weakly constricted posterior to eyes; eyes moderately emarginate, ventral lobe slightly larger than dorsal lobe; epistomal suture weak or obsolete medially; labrum quadrilateral, twice as wide as long, very weakly emarginate and fringed with long, yellow setae anteriorly; mentum nearly planar, about 0.33 times as wide as head, partially concealing ligula and insertions of maxillae; first segment of maxillary palp small, elbowed, partially concealed by mentum; second segment largest, twice length of third, weakly expanded distally; third segment strongly expanded distally, width subequal to length; apical segment about 1.5 times length of third, weakly securiform; labial palpi largely concealed by mentum, segments subcylindrical; antennae subequal to profemur in length, extending posterad slightly beyond middle of pronotum; second segment subquadrate; third segment largest, nearly as long as succeeding two segments combined, weakly expanded distally; segments 4 to 10 gradually expanded into a weak, asymmetrical club; each segment largest medially; terminal segment strongly asymmetrical, distally attenuate.

Pronotum moderately convex, hexagonal or with arcuate margins in dorsal aspect, lateral margins finely beaded, moderately convex; anterior corners broadly rounded, posterior corners angulate, anterior prosternal margin strongly raised; prosternal process prominent and horizontal posteriorly or declivous, subequal to coxa in greatest width, widest between coxae, usually with parallel marginal grooves which continue anterolaterally around coxal cavities.

Elytra strongly convex, broadest near middle, length slightly less than twice breadth; humeri rounded; elytral apex broadly rounded or produced to an acutely rounded angle; epipleuron narrow, average width less than or subequal to width of scutellum, abruptly narrowed at anterior margin of fifth abdominal sternite, gradually widened near humerus; scutellum about 3.0 times as broad as long, lateral margins straight or weakly arcuate.

Mesosternum anteriorly excavated in a broadly rounded notch between mesocoxae; mesotrocantin large; mesepisternum at coxal cavities subequal to diameter of mesocoxal cavity; metasternal suture extending 0.75 times the length of metasternum; metepisternum about 5.0 times longer than broad; metacoxal cavities transverse. Abdomen weakly concave in lateral aspect; abdominal sternites strongly, transversely convex; posterior margins of sternites 3 and 4 curving abruptly posterad laterally; sternite 5 broadly or acutely rounded apically.

Femora subovate in transverse section, weakly to moderately clavate, mesofemur about 1.1 times longer than profemur, metafemur about 1.2 times longer than profemur; tibiae subcylindric in transverse section, subequal in length to corresponding femora; protibia weakly arcuate, metatibia nearly straight; all tibiae with two narrow, parallel lines of fine, yellow decumbent setae along distal half to two-thirds of medial surface; tibial spurs short, barely extending beyond tibial apex; tarsi about half as long as tibiae, proximal four tarsomeres densely covered by short, fine, yellow, decumbent setae, distal tarsomere with two narrow, parallel lines of setae; tarsal claws about half as long as last tarsal segment.

Female.—Similar to male except in following characters: average body size 5 to 10 percent larger than male, slightly more robust; abdomen weakly convex or flat in lateral aspect; legs approximately 10 percent shorter than in male in relation to body size.

KEY TO THE ADULTS OF THE SPECIES OF COELOCNEMIS

1. Hypomeron smooth; punctures obsolete or very fine, sparse, separated by two or three times diameter; subocular rugosity not extending to midventral line of head 2
   Hypomeron distinctly punctate or punctato-rugose, punctures separated by one to two times diameter; subocular rugosity extending as continuous band across postoral region 4
2. Elytral intervals planar 3
   Elytra sulcate ....*sulcata* (p. 90)
3. Prosternal process prominent, acutely angulate; mentum wider than long ......*magna* (p. 87)
   Prosternal process declivous, rounded; mentum longer than wide ....*slevini* (p. 86)
4. Pronotum broadest at or behind middle; ratio of greatest elytral width to elytral length greater than 0.6 5
   Pronotum broadest before middle; ratio of greatest elytral width to elytral length less than 0.6 ....*punctata* (p. 98)

5. Pronotum and cranium very coarsely, densely, usually reticulately punctate..............
  rugulosa, new species (p. 97)
  Pronotum and cranium finely to coarsely punctate, but never reticulately so 6
6. Prosternal process prominent, acutely angulate (endemic to southern California mountains)
  californica (in part) (p. 93)
  Prosternal process declivous, obtusely rounded 7
7. Elytral length 10 to 13.5 mm (endemic to Santa Lucia Mountains, Monterey County,
  California)...*lucia*, new species (p. 96)
  Elytral length 13 to 20 mm; widespread ....*californica* (in part) (p. 93)

Coelocnemis slevini Blaisdell

(Pl. 2, fig. 66)

Coelocnemis slevini Blaisdell, 1925, Proc. Calif. Acad. Sci., 14:337; Blaisdell, 1943, Proc. Calif. Acad. Sci., 24:273.

Male.—Moderate in size, moderately robust, uniformly black, punctation and rugosity greatly reduced, nearly glabrous and moderately lustrous throughout most of body.

Head very finely and obscurely, moderately densely punctate on vertex and frons; postgenae moderately coarsely, asperately and densely punctate posterad of ocular constriction and laterad of gular sutures, weakly rugose posterad of insertion of mouthparts; supraorbital ridges smooth, rounded, subadjacent to eye; mentum barely longer than wide, apex wider than base, moderately cleft and depressed anteriorly, depressed posterolaterally, finely, moderately densely punctate; mandible finely, moderately densely punctate on outer surface

Pronotum 1.25 times wider than long, lateral margins evenly arcuate except just before very weakly exserted posterior angles; posterior angles obtuse to nearly right angled; anterolateral, lateral and posterior margins narrowly, weakly upturned, subequal in width; disk evenly, sparsely and very finely punctate; hypomeron nearly glabrous, a few very fine punctures sometimes visible; prosternum very weakly and smoothly rugose, impunctate or with a few very fine punctures; prosternal process widest anteriorly, gradually narrowed and moderately declivous, then horizontal posterad of coxal cavities, tapering to an acutely rounded apex, glabrous or nearly so.

Elytra moderately convex, abruptly declivous posteriorly, widest at about middle, tapering gradually posterad to broadly rounded apex; strial punctures moderate in size, somewhat irregular, separated by 2.0 to 4.0 times their diameter, sometimes connected by very fine, longitudinal creases, particularly posteriorly; interstriae sparsely, exceedingly finely punctate, very finely, shallowly and confusedly creased; epipleron glabrous or very sparsely, exceedingly finely punctate, finely and shallowly, transversely creased and undulating; scutellum finely, moderately densely punctate, about 3.0 times broader than long, apex obtusely rounded.

Mesosternum and metasternum finely, sparsely, obsoletely punctate, appearing almost glabrous in some specimens; mesopleural and metapleural sclerites moderately coarsely and obsoletely punctate; abdominal sternites 1 through 4 exceedingly finely, sparsely punctate, gently, predominantly transversely undulating, at least in anterior half; fifth sternite finely, densely punctate, submarginally grooved almost to anterior angles, broadly, evenly rounded.

Profemora moderately coarsely, obsoletely and moderately densely punctate, sometimes weakly rugose, especially mesad; mesofemora and metafemora finely, moderately densely punctate; metafemora, and to a less extent mesofemora, finely asperate mesally; tibiae finely, densely punctate.

Female.—Differs from male as stated in generic description.

Measurements.—Elytral length, 14.7 to 17.4 mm; elytral width, 8.7 to 10.2 mm; pronotal width, 7.0 to 8.8 mm; pronotal length, 5.9 to 6.9 mm.

Types.—Holotype male, Grand Canyon, Cedros Island, Baja California, VIII-7-1922 (G. D. Hanna and J. R. Slevin); allotype female, same data as holotype.

The holotype, allotype and paratypes are deposited at the California Academy of Sciences, San Francisco. The holotype is a typical specimen: elytral length, 17.7 mm; elytral width, 10.6 mm; pronotal width, 8.8 mm; pronotal length, 6.9 mm.

C. slevini is similar to specimens of *magna* LeConte from the Sierra San Pedro Martir, Arizona, and southern California, but is differentiated by the elongate mentum and nearly glabrous cuticular surfaces.

Material examined—6 ♂ ♂, 5 ♀ ♀ (see fig. 49). Mexico. Baja California Norte: Cedros Island, VIII-7-1922; Cedros Island, Grand Canyon, VI-5-1925.

Coelocnemis magna LeConte

(Pl. 1, figs. 54–58)

Coelocnemis magna LeConte, 1851, Ann. Lyceum Nat. Hist. N.Y., 5:150; Casey, 1924, Memoirs on the Coleoptera, 11:317.

Coelocnemis obesa LeConte, 1851, Ann. Lyceum Nat. Hist. N.Y., 5:150; Casey, 1924, Memoirs on the Coleoptera, 11:317. New synonymy.

Coelocnemis caudicalis Casey, 1924, Memoirs on the Coleoptera, 11:316. New synonymy.

Coelocnemis deserta Casey, 1924, Memoirs on the Coleoptera, 11:316. New synonymy.

Coelocnemis antennalis Casey, 1924, Memoirs on the Coleoptera, 11:317. New synonymy.

Coelocnemis aequalis Casey, 1924, Memoirs on the Coleoptera, 11:318. New synonymy.

Coelocnemis smithiana Casey, 1924, Memoirs on the Coleoptera, 11:318. New synonymy.

Coelocnemis rotundicollis Casey, 1924, Memoirs on the Coleoptera, 11:319. New synonymy.

Coleocnemis longicollis Casey, 1924, Memoirs on the Coleoptera, 11:319. New synonymy.

Male.—Robust to moderately slender, moderate to large in size, uniformly black, punctation mostly fine, rugosity reduced except in postoral area.

Vertex and frons finely, distinctly, moderately densely punctate, postgenae and amplected portion of vertex coarsely, densely punctate, moderately asperate laterad of gula, postoral area moderately rugose laterally to ventral margins of eyes; supraorbital ridge low, rounded, subadjacent to eye, sometimes separated from eye by shallow groove; mentum about 1.66 times wider than long, apex and base subequal in width, moderately densely, moderately coarsely punctate, anterior border very weakly emarginate and weakly depressed medially, usually weakly to moderately elevated along medial longitudinal ridge and lateral margins; mandibles finely, densely punctate, occasionally weakly rugose on outer surface.

Pronotum about 0.15 to 0.33 times wider than long, lateral margins nearly evenly rounded to broadly explanate and straight or slightly convex behind the middle; posterior angles slightly acute and weakly exserted or nearly right angled; posterior margin broadly raised; disk finely to very finely, evenly, and moderately densely punctate, sometimes subrugose laterally and posteriorly; hypomeron finely, sparsely punctate and subrugose to glabrous; prosternum moderately to very weakly rugose, sometimes finely, sparsely and irregularly punctate; prosternal process prominent, flat, prolonged horizontally posterad of coxal cavities or less commonly moderately declivous, tapering gradually to acutely rounded, or more rarely, obtusely rounded apex, finely to very finely, moderately densely punctate, or nearly glabrous.

Elytra ovate in dorsal aspect, abruptly or gradually declivous posteriorly, widest at or slightly behind middle, tapering to acutely or broadly rounded and frequently weakly appendiculate apex; strial punctures moderate in size, somewhat irregular, separated by two to four times their diameter, usually connected by fine, longitudinal creases; interstriae finely to very finely, moderately densely punctate and finely, shallowly, confusedly creased, weakly undulating or glabrous; epipleuron finely, sparsely punctate to nearly impunctate and coarsely to finely creased transversely or weakly undulating; scutellum 2 to 2.5 times wider than long, finely, irregularly punctate, apex obtusely angulate or broadly rounded.

Mesosternum and metasternum finely, moderately densely punctate and weakly subrugose to very finely, moderately densely punctate and non-rugose; mesopleural and metapleural sclerites densely and moderately coarsely to coarsely punctate, sometimes obscurely so; abdominal sterna 1 through 3 very weakly to moderately, predominantly longitudinally rugose and finely, moderately densely punctate; sternites 4 and 5 scarcely or not rugose, finely, densely punctate; sternite 5 slightly less to slightly more than twice as broad as long, bordered posteriorly by an incomplete submarginal groove.

Profemora moderately densely and moderately coarsely to coarsely punctate; weakly subrugose to moderately rugose, especially basally; mesofemora and metafemora as profemora, but finely, asperately roughened mesally, especially in metafemora; tibiae densely, moderately coarsely punctate.

Female.—Differs from male as stated in generic description.

Types.—Holotype female, deposited at Museum of Comparative Zoology, Harvard University; no collection data. Allotype not designated.

The holotype is similar to specimens from northern California. Measurements are: elytral length, 19.3 mm; elytral width, 9.8 mm; pronotal width, 8.3 mm; pronotal length, 6.9 mm.

Type locality.—Of *magna,* San Jose (Santa Clara County, California); *obesa,* Santa Ysabel (San Diego County, California); *caudicallis,* coast region of central California; *deserta,* southern Arizona; *antennalis,* southern California; *aequalis,* North Fork, Madera County, California; *smithiana,* Milpitas, Santa Clara County, California; *rotundicollis,* Williams, Coconino County, Arizona; *longicollis,* southern Arizona.

C. magna is most similar to *slevini* Blaisdell, from which it is differentiated by the broader mentum and more coarsely punctate cuticle. It is also similar in most characters to *sulcata* Casey, especially individuals from the Providence Mountains, but always has flat elytral intervals.

Extensive populations occur in oak-woodland habitats in California and montane Arizona, with a number of small weakly differentiated populations relictual in desert mountain ranges. Rather than designate each of these as separate taxa all are included in a single polytypic species which is characterized as follows:

Montane Arizona populations (pl. 1, fig. 54).—Moderate in size; punctation very fine, body sometimes nearly impunctate; gula and prosternum very weakly rugose; pronotum about 1.1 to 1.2 times broader than long, lateral pronotal margins nearly evenly rounded, becoming almost straight just before basal angles; elytra 1.6 to 1.8 times longer than broad, evenly rounded and never appendiculate apically; fifth abdominal sternite about 0.4 times as long as broad.

Measurements.—Elytral length, 11.7 to 18.0 mm; elytral width, 6.9 to 11.9 mm; pronotal width, 5.4 to 8.3 mm; pronotal length, 4.8 to 7.2 mm.

This series corresponds to *rotundicollis* Casey and *longicollis* Casey. Although most individuals are separable from cismontane California specimens, the isolated populations in transmontane California are intermediate, suggesting that all should be included in a single species. Collected specimens are predominantly from oak-conifer woodland at elevations from about 1,200 m to 2,400 m.

Range (fig. 48).—Fredonia and the Hualapai Mountains in northern and western Arizona, south to the Baboquiviri, Huachuca, and Chiricahua Mountains, and east to the Sandia and Florida Mountains in New Mexico.

Material examined.—247 ♂ ♂, 199 ♀ ♀.

Owens Valley populations.—Intermediate in most characters between northern California and Arizona populations. Moderate to occasionally large in size; punctation fine to very fine; gula and prosternum weakly rugose; pronotum about 1.1 to 1.2 times broader than long; lateral pronotal margins nearly evenly rounded, straight just before basal pronotal angles. Elytra about 1.8 times longer than broad, evenly rounded, infrequently weakly appendiculate; fifth abdominal sternite about 0.4 times as long as broad.

Measurements.—Elytral length, 12.7 to 19.5 mm; elytral width, 6.9 to 10.7 mm; pronotal width, 5.7 to 8.8 mm; pronotal length, 5.2 to 7.4 mm.

The specimens of this series show high similarities to individuals from the Providence Mountains, Little San Bernardino Mountains, and Arizona as well as from northern California. The habitat occupied by these beetles is semiarid to arid desert scrub and pinyon-juniper woodland, ranging in elevation from about 500 m to 1,500 m.

Material examined.—21 ♂ ♂, 19 ♀ ♀ (see fig. 48). California. Inyo County: Argus Mtns.,

VI-1-1941; IV-1891; Coso Valley, V-1-1941; Harrisburg, Panamint Mtns., V-5-1932, 5,000'; Inyo County, IV-14-1940; Lone Pine, VI-14-1937; Lone Pine, Diaz Lake, V-19-1937; Lone Pine, Goodale Creek, IV-3-1953, on *Salix exigua* flowers; Lone Pine, 2 mi. SW, V-9-1969; Lone Pine, 5 mi. W, VIII-1-1962; Lone Pine, 9 mi. W, VII-7-1961; Mazourka Cyn., 14 mi. NE Independence, V-10-1969; Panamint Springs, 8 mi. S, V-13-1969; Panamint Valley, IV-1891; Saline Valley, Grapevine Cyn., IV-11-1959; IV-2/IX-17-1960; Taboose Camp, VIII-12-1965; Wildrose Cyn., Panamint Mtns., VI-7-1956.

Little San Bernardino Mountains populations.—Intermediate between Arizona and southern California populations. Moderate in size; punctation fine; gula and prosternum weakly to very weakly rugose; pronotum about 1.1 to 1.15 times broader than long, lateral pronotal margins nearly evenly rounded, weakly explanate, straight just before basal angles; elytra 1.6 to 1.8 times longer than broad, evenly rounded, never appendiculate; fifth abdominal sternite about 0.4 times as long as broad.

Measurements.—Elytral length, 14.5 to 16.7 mm; elytral width, 8.4 to 9.4 mm; pronotal width, 6.5 to 7.5 mm; pronotal length, 5.7 to 6.8 mm.

These beetles are similar to the Owens Valley and southern California series, as well as to the Providence Mountains population of *sulcata* Casey. Possibly they were isolated from San Bernardino Mountain populations after the last glaciation.

Material examined.—7 ♂ ♂, 2 ♀ ♀ (see fig. 48). California. San Bernardino County: Joshua Tree Nat. Monument, Covington Flat, VII-18-1965; Rose Mtn., nr. Round Meadow, V-14-1966; 29 Palms, VI-15-1938.

Sierra San Pedro Martir Mountain populations (pl. 1, fig. 55).—Moderate in size, punctation fine to moderately coarse; gula and prosternum weakly to moderately rugose; pronotum about 1.33 times broader than long; pronotal sides weakly explanate, nearly evenly rounded, becoming straight just before basal angles; elytra about 1.6 times longer than broad, evenly rounded, never appendiculate apically; fifth abdominal sternite about 0.4 times as long as broad.

Measurements.—Elytral length, 13.9 to 15.0 mm; elytral width, 8.5 to 9.4 mm; pronotal width, 6.7 to 7.7 mm; pronotal length, 5.3 to 6.0 mm

This series shows similarities to specimens from both Arizona and southern California. Although collection records from the intervening area are lacking, the weak morphological differentiation suggests that the Martir range populations have probably been allopatric with those in southern California since at least the last glaciation.

Material examined.—3 ♂ ♂, 3 ♀ ♀ (see fig. 48). Mexico. Baja California Norte: Sierra Martir, la Encantada, VI-1-1958, 7,000'; la Grulla, V-28-1958, 6,500'; la Sanja, V-28-1958, 6,500'.

Cismontane southern California populations (pl. 1, fig. 56).—Moderate to large in size; punctation fine, occasionally moderately coarse; gula and prosternum moderately to weakly rugose. Pronotum about 1.33 times broader than long, sides of pronotum moderately explanate, evenly, convexly arcuate or nearly straight before middle, evenly, concavely arcuate or straight from middle to basal pronotal angles; elytra 1.6 to 1.8 times longer than wide, moderately attenuate and frequently appendiculate apically; fifth abdominal sternite about 0.5 times as long as broad.

Measurements.—Elytral length, 12.4 to 21.2 mm; elytral width, 7.2 to 12.6 mm; pronotal width, 6.0 to 11.2 mm; pronotal length, 4.7 to 7.8 mm.

The southern California series corresponds to *obesa* LeConte, but shows continuous clinal variation with specimens from northern California. A weak discordance occurs just north of the Transverse Ranges which are arbitrarily considered the northern boundary of this series. The slightly differentiated population in the Mojave River Valley may date from historical times and is included in the southern California series. The habitat is commonly oak woodland and oak-conifer woodland at elevations up to 3,400 m, but specimens have been collected from coastal chaparral near sea level.

Range (fig. 48).—Throughout montane and cismontane southern California and western Los Angeles County in the northwest, east to the Antelope Valley (Palmdale, Little Rock) and Mojave River Valley (Victorville and Yermo), south along the eastern escarpment of the Peninsular Ranges (Indio, Chino Cyn., Palm Cyn., Palm Springs, Warner Springs) and west of the Peninsula ranges to the vicinity of Ensenada in Baja California.

Material examined.—560 ♂ ♂, 665 ♀ ♀.

Cismontane northern California populations (pl. 1, fig. 57).—Moderate to large in size, punctation fine, sometimes moderately coarse; gula and prosternum moderately to weakly rugose. Pronotum about 1.2 times broader than long; lateral pronotal margins evenly arcuate or nearly straight from middle to basal pronotal angles; elytra 1.8 to 1.9 times longer than broad, tapering to acutely rounded apex, frequently weakly appendiculate apically; fifth sternite about 0.5 times as long as broad.

Measurements.—Elytral length, 13.4 to 22.1 mm; elytral width, 7.4 to 12.0 mm; pronotal width, 5.6 to 10.2 mm; pronotal length, 4.8 to 8.1 mm.

These specimens are differentiated from those of southern California by the more slender body, but integradation occurs in the southern Coast Ranges and southern Sierra Nevada. This series corresponds to *magna* LeConte. The known habitat includes chaparral and oak woodland or oak-conifer woodland at elevations from near sea level to more than 1,500 m.

Range (fig. 48).—Cismontane California, from interior Mendocino County (Covelo) and Shasta County (Redding), south through the interior Coast Ranges and Sierra Nevada to southern Kern and San Luis Obispo counties.

Material examined.—435 ♂ ♂, 387 ♀ ♀.

Santa Catalina Island populations (pl. 1, fig. 58).—Very similar to mainland southern California populations in most characters other than cuticular sculpturing. Moderate in size, punctation moderately coarse; elytra, gula and prosternum moderately rugose; pronotum about 1.15 to 1.33 times broader than long; lateral pronotal margins moderately explanate, evenly rounded or nearly straight before middle, evenly, concavely arcuate or straight from middle to basal pronotal angles; elytra about 1.6 times longer than broad, moderately attenuate and sometimes appendiculate apically; fifth abdominal sternite about 0.5 times as long as broad.

Measurements.—Elytral length, 15.2 to 17.8 mm; elytral width, 9.6 to 10.6 mm; pronotal width, 7.1 to 8.5 mm; pronotal length, 5.5 to 6.8 mm.

These specimens occur in localized oak woodland in mesic canyon bottoms. They are clearly derived from the southern California mainland, but the date at which Santa Catalina Island was invaded is unknown.

Material examined.—4 ♀ ♀ (see fig. 48).—United States. California. Los Angeles County: Santa Catalina Island, Avalon, IV-20; Middle Canyon, VI-18-1969.

Coelocnemis sulcata Casey
(Pl. 1, figs. 59–61)

Coelocnemis sulcata Casey, 1895, Coleopterological Notices, 5:615.

Male.—General form moderately slender to robust, moderate to large in size, uniformly black to reddish black, finely to very finely punctate, elytra coarsely sulcate.

Vertex and frons distinctly, finely to very finely, moderately densely punctate; amplected portions of cranium moderately coarsely, densely, asperately punctate; postgenae coarsely, densely punctate; postoral region moderately rugose laterally nearly to ventral margin of eyes or weakly rugose; clypeus finely to very finely, moderately densely punctate centrally, more finely, densely punctate along anterior margin; epistomal suture well defined, at least laterally;

gula impunctate, with fine intersecting lines basally; supraorbital ridge raised, prominent, or low, rounded, subadjacent to eye, separated from it posteriorly by a shallow groove; mentum about 1.5 times wider than long, densely, irregularly and moderately to coarsely punctate, anterior and posterior edges subequal in width, anterior border very weakly emarginate or straight, very weakly depressed medially; nearly planar centrally or with raised lateral margins and median longitudinal ridge or groove; mandibles moderately coarsely, densely punctate to weakly punctato-rugose.

Pronotum moderately convex, 1.05 to 1.20 times broader than long, widest at about middle, lateral margins evenly rounded to straight or slightly concave behind the middle; posterior angles right angled or slightly obtuse, sometimes weakly exserted; lateral margins narrowly, weakly upturned, posterior margin broadly raised, finely, densely punctate; disk finely to very finely, moderately densely punctate; hypomeron very finely, obsoletely punctate, to finely but distinctly, sparsely punctate, smooth to undulating, subrugose; prosternum glabrous to finely, sparsely, irregularly punctate, subrugose to moderately rugose; prosternal process prominent, horizontal behind coxae, tapering to an acutely rounded apex, lateral borders submarginally grooved and steeply upturned between coxal cavities, finely, sparsely to moderately densely pnnctate.

Elytra moderately to broadly convex, moderately declivous posteriorly; widest at middle, tapering to a broadly, acutely rounded apex, never appendiculate; strial punctures moderate in size, irregular, separated by 1.0 to 5.0 times diameter, distinct or obscured by broad, shallow creases running transversely or obliquely through them; interstriae finely to very finely, moderately densely punctate, smooth, finely to moderately, confusedly creased or finely subrugose, moderately to strongly convex; epipleura glabrous to finely, sparsely punctate, crossed by numerous fine to moderately coarse, transverse creases, to slightly undulating; scutellum about 2.5 to 3.0 times wider than long, finely, moderately densely, irregularly punctate.

Mesosternum finely, moderately densely to sparsely punctate medially, moderately coarsely, moderately densely punctate laterally; metasternum finely to very finely, moderately densely punctate medially, sometimes with a few coarse punctures laterally, smooth to subrugose; metapleural sclerites moderately coarsely, moderately densely punctate; abdominal sternites 1 to 3 finely to very finely, moderately densely punctate, finely to moderately, predominantly longitudinally, rugose, at least in anterior half; sternites 4 and 5 finely, moderately densely to densely punctate; sternite 5 bordered by an incomplete, submarginal groove; 0.4 to 0.5 times as long as broad.

Profemora moderately densely and moderately coarsely punctate anteriorly to coarsely punctate posteriorly; sometimes weakly subrugose posterobasally; mesofemora and metafemora as profemora, but finely, asperately roughened mesally, especially in metafemora; tibiae densely, moderately coarsely punctate.

Female.—Differs from male as stated in generic description.

Types.—Holotype female, Southwestern Utah (C. J. Weidt); allotype not designated.

The holotype is deposited at the United States National Museum. Casey's type series apparently included one other female, now bearing a paratype label. The holotype is a specimen similar to others from the Great Basin: elytral length, 19.3 mm; elytral width, 10.4 mm; pronotal width, 8.1 mm; pronotal length, 7.1 mm.

C. sulcata is similar to *magna* LeConte in most characters, but is readily separated by the sulcate elytral intervals. The species as considered here includes three allopatric series of populations of distinct phenotype, with no intermediates known.

Great Basin populations (pl. 1, fig. 59).—Moderate to large in size; black in color; punctation fine to moderately coarse; prosternum and gula moderately to weakly rugose; prosternal process acutely angulate posteriorly, usually prominent, horizontal or occasionally weakly declivous; elytra usually strongly sulcate, occasionally moderately or weakly so, about 1.7 to 1.9 times longer than broad.

Measurements.—Elytral length, 12.1 to 20.0 mm; elytral width, 6.7 to 11.2 mm; pronotal width, 5.2 to 8.8 mm; pronotal length, 4.8 to 7.5 mm.

Individuals of this series are quite similar throughout their range, but those from Coral Pink Sand Dunes, Kane County, Utah, sometimes have reddish black elytra. Casey (1895) reports that *C. sulcata* occupies lower, more arid habitats than *punctata* LeConte, but personal observations reveal that they occur in at least partial sympatry in pinyon-juniper woodland.

Range (fig. 49).—Montane regions of the Great Basin, from southern Nevada (Charleston Mountains) and Utah (St. George), north to Reno, Nevada, and southeastern Idaho (Idaho Falls, Soda Springs).

Material examined.—92 ♂ ♂, 85 ♀ ♀.

Montane Arizona populations (pl. 1, fig. 60).—Moderate in size; black or occasionally reddish black in color; punctation fine to very fine; prosternum and gula very weakly rugose; prosternal process always prominent, horizontal, acutely angulate posteriorly; elytra moderately or occasionally weakly or strongly sulcate, about 1.6 to 1.8 times longer than broad.

Measurements.—Elytral length, 12.5 to 18.6 mm; elytral width, 7.1 to 10.5 mm; pronotal width, 5.4 to 8.2 mm; pronotal length, 5.0 to 7.4 mm.

This series includes specimens from central and southern Arizona, separated from the Great Basin populations by a broad distributional gap in northwestern Arizona. Phenetically the Arizona populations resemble Arizona populations of *magna* LeConte in characters other than elytral sulcation, suggesting that *sulcata* may be polyphyletic as constituted here. It is also possible that the more rotund body of the Arizona populations represents an adaptive trend similar to that in cismontane California populations of *magna*, where the more rotund individuals occur in the southern portion of the range. Specimens from the Yavapai County localities tend to be reddish black in coloration, but the Arizona populations are otherwise homogeneous. The habitat is dry pine–oak woodland from about 1,500 m to 2,000 m, but sparser populations also occur in the surrounding chaparral at slightly lower elevations.

Range (fig. 49).—Montane Arizona, from Coconino and Yavapai counties in the north, south to Pima County and east to southern Navajo County.

Material examined.—96 ♂ ♂, 82 ♀ ♀.

Providence Mountains populations (pl. 1, fig. 61).—Moderate in size, black in color; punctation fine to very fine, prosternum and gula weakly to very weakly rugose; prosternal process acutely angulate posteriorly, prominent and horizontal or occasionally weakly declivous; elytra weakly to moderately sulcate, about 1.8 times longer than broad.

Measurements.—Elytral length, 14.0 to 18.4 mm; elytral width, 8.0 to 10.4 mm; pronotal width, 6.2 to 8.2 mm; pronotal length, 5.8 to 7.6 mm.

The characteristically weakly sulcate elytra are also seen in a few specimens from the White Mountains, Inyo County, which are included in this series. The habitat is pinyon-juniper woodland above 1,200 m elevation.

Material examined.—29 ♂ ♂, 25 ♀ ♀ (see fig. 49). California. Inyo County; Westgard Pass, 3.6 mi. E, VII-10-1967; Wyman Cyn., VII-10-1967. San Bernardino County: Providence Mtns., Cedar Cyn., VI-10-1940, V-11-1969, VII-1966, 5,050–5,500', VIII-1966, IX-10-1966; Providence Mtns., IV-17-1965, 4,500', under pinyon, IV-10-1949.

Coelocnemis californica Mannerheim
(Pl. 1, fig. 62; Pl. 2, figs. 63–65)

Coelocnemis californica Mannerheim, 1843, Bull. Soc. Imp. Nat. Moscou, 16:282; Mannerheim, 1844, Mag. Zool., Vol. 6 (Ser. 2), Part 133, p. 2; Casey, 1924, Memoirs on the Coleoptera, 11:315.

Coelocnemis dilaticollis Mannerheim, 1843, Bull. Soc. Imp. Nat. Moscou, 16:282; Mannerheim, 1844, Mag. Zool., Vol. 6 (Ser. 2), Part 133, p. 3; Casey, 1924 Memoirs on the Coleoptera, 11:315. New synonymy.

Coelocnemis colombiana Casey, 1924, Memoirs on the Coleoptera, 11:314. New synonymy.

Coelocnemis rauca Casey, 1924, Memoirs on the Coeloptera, 11:314. New synonymy.

Coelocnemis ovipennis Casey, 1924, Memoirs on the Coleoptera, 11:315. New synonymy.

Coelocnemis utensis Casey, 1924, Memoirs on the Coleoptera, 11:315. New synonymy.

Coelocnemis spaldingi Casey, 1924, Memoirs on the Coleoptera, 11:316. New synonymy.

Coelocnemis basalis Casey, 1924, Memoirs on the Coleoptera, 11:316. New synonymy.

Coelocnemis idahoensis Casey, 1924, Memoirs on the Coleoptera, 11:317. New synonymy.

Coelocnemis rugosa Linnell, 1901, Proc. Entomol. Soc. Wash., 4:181. New synonymy.

Coelocnemis barretti Blaisdel, 1928, Pan-Pac. Entomol., 4:163. New synonymy.

Male.—Moderately to very robust, small to large in size, uniformly black, finely to coarsely punctate, finely to very coarsely rugose.

Vertex and frons distinctly, coarsely and moderately densely to densely punctate to punctato-rugose or rugose; amplected portion of cranium asperately, coarsely, very densely punctate; postgenae and postoral region very coarsely, sharply rugose laterally nearly to dorsal margin of eyes; clypeus coarsely, densely punctate to coarsely punctato-rugose centrally, more finely, very densely punctate along anterior margin; epistomal suture well defined or obscured by punctation; gula impunctate, with a few fine, intersecting lines basally; supraorbital ridge raised, prominent, or low, rounded, subadjacent to eye, separated from it posteriorly by a shallow groove; mentum about 1.5 times wider than long, anterior and posterior edges subequal in width, moderately densely, moderately coarsely punctate to moderately rugose, anterior border scarcely emarginate, very weakly depressed medially, nearly planar centrally, or with raised lateral margins and medial longitudinal ridge or groove; mandibles densely, moderately coarsely punctate to punctato-rugose on outer surface.

Pronotum moderately convex, 1.25 to 1.53 times wider than long, widest at or slightly behind middle, lateral margins narrowly, weakly beaded, nearly evenly rounded to broadly, sometimes angulately explanate, straight or evenly convex before the middle, moderately convex to nearly straight behind the middle; posterior angles very weakly to strongly exserted, slightly obtuse to slightly acute; posterior margin broadly raised, finely, densely punctate; disk moderately densely and moderately coarsely to coarsely punctate to coarsely punctato-rugose along posterior and lateral margins; hypomeron moderately coarsely, moderately densely, sometimes obscurely punctate to very coarsely, densely punctate or very coarsely punctato-rugose, especially posteriorly and over coxae; prosternum weakly to moderately rugose or punctato-rugose to strongly, coarsely rugose, punctation usually obscured by rugosity; prosternal process strongly declivous, flat and broadly rounded behind coxae, moderately declivous and rounded, or prominent, horizontal, tapering to an acutely rounded apex; lateral borders shallowly grooved submarginally and moderately upturned between coxal cavities; moderately densely and moderately coarsely punctate to punctato-rugose.

Elytra broadly convex, abruptly or moderately declivous posteriorly, widest at middle, tapering to a broadly rounded apex, never appendiculate; strial punctures moderate in size, irregular, separated by 1.0 to 5.0 times diameter, usually connected by fine, longitudinal creases, the latter variously obscured by rugosity, completely obliterated in most rugose specimens; interstriae planar, moderately densely and finely or moderately coarsely punctate and finely, shallowly, confusedly creased or subrugose to coarsely, densely punctate and coarsely rugose or very coarsely rugose, with the punctation nearly obliterated; epipleura sparsely to moderately densely, finely punctate, crossed by a few to numerous fine to moderately coarse, transverse creases or slightly undulating; scutellum 2.0 to 2.5 times wider than long, finely to moderately coarsely and densely but irregularly punctate, apex nearly right angled or obtusely angulate.

Mesosternum finely to moderately coarsely and sparsely punctate medially, coarsely, densely punctate laterally; metasternum moderately coarsely and densely punctate medially with a few coarse lateral punctae to coarsely, moderately densely punctate and moderately rugose; pleural sclerites coarsely and sparsely to densely punctate; abdominal sternites 1 to 3 densely, moderately coarsely to very coarsely punctate, to coarsely punctato-rugose; sternites 1, 2 and usually 3 weakly to moderately, predominantly longitudinally rugose, at least in anterior half; sternites 4 and 5 moderately coarsely, densely punctate to coarsely, moderately densely punctate or punctato-rugose; sternite 5 sometimes bordered posteriorly by an incomplete, submarginal groove; 0.33 to 0.40 times as long as broad.

Profemora coarsely, densely punctate anteriorly and distally to coarsely punctato-rugose posteriorly and proximally; mesofemora and metafemora as profemora, but finely, asperately roughened mesally, especially in metafemora; tibiae densely, regularly, moderately coarsely punctate.

Female.—Differs from male as stated in the generic description.

Types.—Holotype not designated. The specimens originally collected by Blaschke were placed in the Obert collection, St. Petersburg. Their present location is not known.

Type locality.—Of *californica,* 'New California;' *dilaticollis,* 'New California;' *colombiana,* British Columbia, Canada; *rauca,* California; *ovipennis,* southern California; *utensis,* Provo Canyon, Utah; *spaldingi,* Provo Canyon, Utah; *basalis,* California; *idahoensis,* Coeur d'Alene, Idaho; *rugosa,* Los Angeles County, California; *barretti,* Bear Canyon, Sierra Madre Mountains, Los Angeles County, California.

C. californica is similar to *punctata* LeConte, from which it is differentiated by its stouter body. *Californica* differs from *rugulosa* n. sp. in its larger body size and finer, non-reticulate cranial and pronotal punctation (figs. 73, 74). In this species are placed individuals from a broad spectrum of localities ranging from southern California to British Columbia and south through Idaho to Utah. Various geographic segregates are well differentiated from one another, but nearly all of the variation is gradual. The series of populations characterized here includes the extremes of variation; the characters and geographic areas in which intermediacy occur are listed for each series.

Pacific Northwest populations (pl. 1, fig. 62).—Moderate to large in size; moderately robust; pronotum and head moderately coarsely punctate, subrugose; prosternal process declivous and flattened, rounded or truncate posteriorly; elytra moderately rugose or occasionally subrugose, coarsely punctate, evenly convex in lateral aspect.

Measurements.—Elytral length, 11.8 to 19.1 mm; elytral width, 7.4 to 12.1 mm; pronotal width, 6.0 to 9.7 mm; pronotal length, 4.8 to 7.0 mm.

This series contains specimens from most of the range of the species, including Utah, Idaho, British Columbia, Washington, Oregon, and California north of the Sierra Nevada and the Russian River. The beetles are morphologically homogeneous throughout this area except at its southern limits in California, where intermediacy with the San Francisco Bay region and Sierran phenotypes occurs.

The commonest habitat is western conifer forest or conifer woodland. Individuals have been taken from near sea level in north coastal California, Washington, and British Columbia and range up to at least 2,000 m elevation in Utah and California.

This series includes *dilaticollis* Mannerheim and most of the species described by Casey (1924).

Range (fig. 50).—Southern British Columbia and Alberta, Canada, south through Idaho, Western Montana, Wyoming, and the Wasatch Mountains in Utah to the vicinity of St. George; south through the montane portions of Washington, Oregon, and California to southern Mendocino, Shasta, and Lassen Counties.

Material examined.—584 ♂ ♂, 523 ♀ ♀.

Sierra Nevada populations (pl. 2, fig. 63).—Moderate to large in size; moderately robust; pronotum and head coarsely punctate, sometimes weakly rugose; prosternal process declivous and flattened; rounded or truncate posteriorly or prominent, horizontal; acutely prominent posteriorly in some individuals from Kern and Tulare counties; elytra strongly, coarsely rugose, rugosity partially obscuring coarse punctation, frequently flattened dorsally in lateral aspect.

Measurements.—Elytral length, 12.9 to 18.6 mm; elytral width, 7.5 to 11.6 mm; pronotal width, 6.3 to 9.8 mm; pronotal length, 4.8 to 7.0 mm.

Occasional specimens with pronounced dorsal elytral flattening occur in the Pacific Northwest series, with which these specimens gradually intergrade in the north. To the south body size gradually decreases, approaching the southern California phenotype. Some specimens from Kern and Tulare Counties also show the prominent prosternal process typical of southern California beetles. The Sierra Nevada populations occur in oak-conifer woodland from about 900 m to 2,500 m elevation.

Range (fig. 50).—South of the Feather River drainage in Plumas County, along the Sierra Nevada Range to the Greenhorn Mountains in Kern County.

Material examined.—342 ♂ ♂, 381 ♀ ♀.

Southern California populations (pl. 2, fig. 64).—Small to moderate in size, pronotum and head coarsely punctate, usually subrugose; prosternal process prominent, horizontal, acutely angulate posteriorly; elytra coarsely rugose, rugosity obscuring coarse punctation, evenly convex in lateral aspect.

Measurements.—Elytral length, 11.1 to 16.2 mm; elytral width, 6.4 to 10.4 mm; pronotal width, 5.5 to 8.4 mm; pronotal length, 4.2 to 6.2 mm.

The specimens comprising this series are differentiated from Sierran individuals primarily by the prominent prosternal process and slightly smaller average size. Included are *ovipennis* Casey, *rugosa* Linnell and *barretti* Blaisdell. The altitudinal range is from about 900 m to more than 3,000 m elevation in oak-conifer and conifer woodland habitats.

Range (fig. 50).—From the Sierra Madre Mountains (Bates Canyon) in Santa Barbara County, east to Mount Pinos, Kern County, then south and east through the Transverse Ranges and Peninsular Ranges to southern San Diego County.

Material examined.—252 ♂ ♂, 285 ♀ ♀.

San Francisco Bay region populations. (pl. 2, fig. 65).—Moderate in size; pronotum and head finely to moderately coarsely punctate, nonrugose; prosternal process strongly declivous and flattened, rounded or truncate posteriorly; elytra alutaceous to subrugose, finely to moderately coarsely punctate, evenly convex in lateral aspect.

Measurements.—Elytral length, 11.6 to 19.1 mm; elytral width, 6.3 to 11.3 mm; pronotal width, 5.8 to 9.6 mm; pronotal length, 4.3 to 7.1 mm.

The smooth, finely punctate sculpturing differentiates these specimens from other series of *C. californica,* but intergradation with the Pacific Northwest series occurs in north coastal California. The habitat of these beetles is oak-conifer woodland in Marin County and the Santa Cruz Mountains and mesic oak woodland at localities that are under 600 m elevation. This series corresponds to *californica* Mannerheim.

Range (fig. 50).—Sonoma and Napa Counties, south through the outer Coast Ranges to northern Monterey County.

Material examined.—87 ♂ ♂, 91 ♀ ♀.

Coelocnemis lucia Doyen, new species

(Pl. 2, fig. 67)

Male.—General form moderately robust, small in size, uniformly black, moderately punctate, weakly to moderately rugose.

Vertex and frons moderately coarsely, moderately densely punctate; postgenae and postoral region of cranium coarsely, sharply rugose laterally nearly to ventral margin of eye; clypeus moderately coarsely and densely punctate laterally, more sparsely so along anterior margin; epistomal suture usually evident, at least in medial portion; supraorbital ridge raised, prominent, subadjacent to eye, separated from it posteriorly by a shallow groove; mentum about 1.5 times wider than long, anterior and posterior edges subequal in length, densely, moderately coarsely punctate and subrugose, anterior border weakly emarginate, surface depressed medially, nearly planar or with raised lateral margins and median longitudinal ridge or groove; outer mandibular surface moderately coarsely and densely punctate to punctato-rugose.

Pronotum moderately, evenly convex, 1.20 to 1.40 times wider than long, widest at or slightly behind middle, lateral margins nearly evenly rounded, or occasionally slightly angulate and nearly straight or weakly concave behind middle; posterior angles weakly to moderately exserted, slightly obtuse to slightly acute; lateral margins narrowly, weakly upturned, posterior margin broadly raised, finely, densely punctate; disk moderately densely, moderately coarsely punctate, usually more coarsely and densely so along posterior and lateral margins; hypomeron moderately coarsely and densely punctate; prosternum moderately punctato-rugose; prosternal process moderately to strongly declivous, rounded behind coxae, lateral borders shallowly grooved submarginally and moderately upturned between coxal cavities; moderately densely, moderately coarsely punctate.

Elytra strongly convex, strongly, but gradually declivous posteriorly, widest at middle, tapering to a broadly rounded apex, never appendiculate; strial punctures moderate in size, separated by 1.0 to 5.0 times diameter; interstriae planar, moderately densely, moderately coarsely punctate, densely subrugose, obscuring punctation; epipleura moderately densely, finely punctate, crossed by moderately coarse transverse creases; scutellum 2.0 to 2.5 times wider than long, finely, irregularly punctate, apex nearly right angled or obtusely angulate.

Thoracic sterna moderately coarsely, sparsely punctate medially, coarsely, densely punctate laterally; pleural sclerites moderately coarsely and densely punctate; abdominal sternites 1 to 3 moderately coarsely punctate, usually more coarsely so laterally, finely, predominantly longitudinally rugose, at least in anterior half; sternites 4 and 5 densely, moderately coarsely punctate; fifth sternite 0.33 to 0.40 times as long as broad, frequently bordered posteriorly by an incomplete submarginal groove.

Profemora moderately coarsely, densely punctate anteriorly and distally to punctato-rugose posteriorly and proximally; mesofemora and metafemora as profemora, but finely, asperately roughened mesally especially in metafemora; tibiae densely, regularly, moderately coarsely punctate.

Female.—Differs from male as stated in generic description

Measurements.—Elytral length, 10.3 to 13.3 mm; elytral width, 6.6 to 8.4 mm; pronotal width, 4.8 to 6.7 mm; pronotal length, 3.8 to 4.9 mm.

Types.—Holotype female, Monterey County, Nacimiento Summit, 5 air mi. E. Lucia, April 20, 1969 (J. T. Doyen); allotype male and 6 male, 10 female paratypes, same data as holotype.

The holotype and allotype are deposited at the California Academy of Sciences, San Francisco, and paratypes at the U.S. National Museum, Washington, D.C., the Canadian National Museum, Ottawa, Ohio State University, Columbus, Ohio, and in the California Insect Survey, Berkeley.

C. lucia is locally common in oak and oak-conifer woodland above 1,000 m in the Santa Lucia Mountains. It is similar in general appearance to *rugulosa* and southern California montane populations of *californica*, differing in the declivous prosternal process (rather than prominent) and in the finer cuticular punctation and rugosity. *C. lucia* differs from northern California and Pacific Northwest populations of *californica* in its much smaller size and stouter body proportions.

Material examined.—17 ♂ ♂, 24 ♀ ♀ (see fig. 51). California. Monterey County: Arroyo Seco Camp, VIII-15-1964; Escondido Camp, 17 air mi. SW King City, IV-19-1969; Jamesburg, 7 mi. S, V-18-1968; Nacimiento Summit, 5 air mi. E. Lucia, IV-20-1969; Tassajara Springs, IX-15-1908.

Coelocnemis rugulosa Doyen, new species

(Pl. 2, figs. 68, 74)

Male.—Stout, robust, small to moderate in size, uniformly black, coarsely to very coarsely punctate, very coarsely rugose.

Vertex and frons coarsely, very densely punctate to coarsely, reticulately punctate; postgenae and postoral region of cranium very coarsely, sharply rugose laterally nearly to ventral margin of eye; clypeus coarsely punctato-rugose centrally, punctation sparser, finer along anterior margin; epistomal suture frequently obscured by punctation; supraorbital ridge raised, prominent, subadjacent to eye, separated from it posteriorly by a shallow groove; mentum about 1.5 times wider than long, anterior and posterior edges subequal in length, densely, moderately coarsely punctate to moderately rugose, anterior border very weakly emarginate, depressed medially, nearly planar centrally, or with raised lateral margins and medial longitudinal ridge or groove; mandibles densely, moderately coarsely punctate to punctato-rugose on outer surface.

Pronotum moderately, evenly convex, 1.20 to 1.45 times wider than long, widest at or slightly behind middle, lateral margins nearly evenly rounded to broadly angulate and straight or evenly convex before the middle, weakly concave to nearly straight behind the middle; posterior angles weakly to strongly exserted, slightly obtuse to slightly acute; lateral margins narrowly, weakly upturned, posterior margin broadly raised, finely, densely punctate; disk coarsely, densely punctate to coarsely, reticulately punctate, especially along posterior and lateral margins; hypomeron coarsely punctato-rugose, especially posteriorly and over coxae; prosternum coarsely rugose, punctation usually obscured by rugosity; prosternal process weakly declivous, rounded posteriorly; moderately densely and coarsely punctate.

Elytra broadly convex, abruptly declivous posteriorly, widest at middle, tapering to a broadly rounded apex, never appendiculate; coarsely to very coarsely rugose; strial punctures and interstrial punctation nearly or quite obliterated by rugosity; interstriae planar; epipleura moderately densely, finely punctate, crossed by few to numerous fine to moderately coarse, transverse creases; scutellum 2.0 to 2.5 times wider than long, moderately coarsely, irregularly punctate; apex right angled to obtusely angulate.

Mesosternum moderately densely punctate and moderately rugose; pleural sclerites coarsely, sparsely to densely punctate, more coarsely so laterally, or coarsely punctato-rugose; sternites 1, 2 and 3 moderately, predominantly longitudinally rugose; sternites 4 and 5 moderately coarsely, densely punctate or punctato-rugose; sternite 5 sometimes incompletely bordered posteriorly by an incomplete, submarginal groove, 0.33 to 0.40 times as long as broad.

Profemora coarsely, densely punctate anteriorly and distally to coarsely punctato-rugose posteriorly and proximally; mesofemora and metafemora as profemora, but finely, asperately roughened mesally, especially in metafemora; tibiae densely, regularly, moderately coarsely punctate.

Female.—Differs from male as stated in generic description.

Measurements.—Elytral length, 10.9 to 14.2 mm; elytral width, 6.9 to 9.2 mm; pronotal width, 5.5 to 7.4 mm; pronotal length, 4.4 to 5.6 mm.

Types.—Holotype, female, California, Modoc County, near Buck Creek Ranger Station, Warner Mountains, VI-9-1970 (R. Dietz and P. Rude); allotype male and 3 male 2 female paratypes, same data as holotype.

The holotype and allotype are deposited at the California Academy of Sciences, San Francisco, and paratypes at the U. S. National Museum, Washington, D.C., and the Canadian National Museum, Ottawa.

This species is similar to *lucia* and the southern California montane populations of *californica* in general appearance. It differs from the former in the moderately prominent prosternal process and from both in the reticulate punctation of the cranium and pronotum (pl. 2, fig. 74). It differs from northern California populations of *californica* in the small size and stout body proportions, as well as its coarser cuticular sculpturing.

C. rugulosa occurs sympatrically with *C. californica* in the dry conifer woodland of the Modoc Plateau in California and Oregon. Specimens have been taken at over 1,800 m elevations in the Warner Mountains and on Mt. Shasta and as low as 1,200 m at the edges of the surrounding basins.

Material examined.—19 ♂ ♂, 20 ♀ ♀ (see fig. 51). California. Lassen County: Facht, VII-17-1921; Grassy Lake, Lassen Nat. Forest, IX-27-1914; Lassen Nat. Park, VII-9-1923; VII-16-1958; Manzanita Lake, Lassen Nat. Park, X-1-1944, 5,800'; Norval Flats, VI-20-1920, 5,500'. Modoc County: Davis Creek, VII-12-1922; Fandango Pass, Warner Mtns., VI-4-1970. Plumas County: Plumas Co., VI-8-1913. Shasta County: Hat Creek, VIII-12-1956; VI-15-1948, under bark; Hat Creek, 9 mi. W, VII-3-1955; Lake Eiler, VII-9-1947; Old Station, VI-24-1955. Siskiyou County: McCloud, X-15-1918; Mt. Shasta, VIII-4-1905, 6,000'; Sand Flat, Mt. Shasta, VII-2-1970; Shasta City, VII; Shasta Springs, VII, VII-6-1904.

Oregon. Deschutes County: Sisters, VII-19-1. Jackson County: Ashland Peak, Siskiyou Mtns., VI-20-1952, 7,000'. Klamath County: Klamath Co., V-19/23-1913.

Coelocnemis punctata LeConte

(Pl. 2, figs. 69–72)

Coelocnemis punctata LeConte, 1854, Proc. Acad. Nat. Sci. Phil., 7(1855):225; Casey, 1924, Memoirs on the Coleoptera, 11:319.

Coelocnemis angusta Casey, 1924, Memoirs on the Coleoptera, 11:319. New synonymy.

Coelocnemis tanneri Blaisdell, 1928, Pan-Pac. Entomol., 4:164. New synonymy.

Male.—Slender, small to moderate in size, uniformly black to reddish black, finely to coarsely punctate, nonrugose to moderately, reticulately rugose.

Vertex and frons moderately densely to densely, moderately coarsely to coarsely, occasionally obscurely punctate; amplected portion of cranium moderately coarsely, very densely punctate; postgenae and postoral region coarsely, sharply rugose becoming coarsely punctato-rugose near dorsal margin of eye; clypeus moderately densely to densely and moderately coarsely to coarsely punctate centrally, more finely and densely punctate along anterior margin; epistomal suture well defined; gula impunctate with a few, fine intersecting lines basally; supraorbital ridge raised, prominent, to low, rounded, subadjacent to eye, separated from it posteriorly by a shallow groove; mentum about 1.5 times broader than long, anterior and posterior edges subequal in width, moderately coarsely and moderately densely to densely punctate, anterior border very weakly emarginate or straight, very weakly depressed medially, nearly planar centrally or with raised lateral margins and median longitudinal ridge; mandibles densely, moderately coarsely punctate to punctato-rugose on outer surface.

Pronotum moderately convex, 0.95 to 1.28 times wider than long, widest at or slightly before middle, lateral margins narrowly, weakly beaded, evenly and weakly to moderately convex, sometimes weakly explanate in middle; posterior angles moderately exserted, near 90° to non-exserted, obtuse; posterior margin broadly raised, finely, irregularly punctate; disk moderately densely to densely and moderately coarsely to coarsely punctate, occasionally subrugose postero-laterally; hypomeron coarsely, moderately densely to densely punctate, smooth to moderately rugose; prosternum weakly to moderately rugose, irregularly, coarsely punctate; prosternal process moderately declivous and acutely to obtusely rounded to strongly declivous and broadly rounded behind coxae; lateral margins deeply grooved submarginally and strongly upturned between coxal cavities; moderately densely, moderately coarsely punctate to punctato-rugose.

Elytra narrowly convex, abruptly or moderately declivous posteriorly, widest at middle, tapering to broadly rounded apex, never appendiculate; strial punctures moderate in size, irregular, separated by 1.0 to 3.0 times diameter, sometimes connected by fine, longitudinal creases or completely obscured by transverse rugae; interstriae finely to moderately coarsely punctate, subrugose, weakly undulating, to coarsely, reticulately rugose, planar to moderately convex; epipleura finely, sparsely punctate to glabrous, crossed by numerous fine to moderately coarse creases or weakly undulating; scutellum 2.0 to 2.25 times as broad as long, finely, irregularly punctate, apex nearly right angled or obtusely angulate.

Mesosternum coarsely, sometimes obscurely, sparsely to moderately densely punctate; abdominal sternites 1 to 3 moderately densely, finely to moderately coarsely punctate, very finely to moderately rugose, at least in anterior half; sternites 4 and 5 finely to moderately coarsely, densely punctate or weakly punctato-rugose; fifth sternite sometimes bordered by an incomplete submarginal groove; about 0.4 times as long as broad.

Profemora moderately coarsely, densely punctate anteriorly and distally to coarsely, densely punctate posteriorly and proximally; mesofemora and metafemora as profemora, but finely, asperately roughened mesally, especially in metafemora; tibiae densely, regularly, finely to moderately coarsely punctate.

Female.—Differs from male as stated in generic description.

Types.—Holotype female deposited at the Museum of Comparative Zoology, Harvard University. Allotype not designated. The holotype is a typical Great Basin specimen of *punctata*.

Measurements.—Elytral length, 13.8 mm; elytral width, 7.8 mm; pronotal width, 6.1 mm; pronotal length, 5.5 mm.

Type locality.—Of *punctata*, along latitude 38°, probably from Utah (Beckwirth); *angusta*, Fort Wingate, McKinley County, New Mexico (Woodgate); *tanneri*, Zion Canyon, Utah, VI-10-1934 (V. M. Tanner).

C. punctata differs from *californica* Mannerheim in its more slender body proportions. The Colorado Plateau specimens are somewhat similar to *sulcata* Casey in elytral sculpturing, but differ in the more coarsely punctate cuticle and strongly rugose postoral region. The phenotypic variation of this species is summarized below.

Great Basin populations (pl. 2, figs. 69, 70, 71).—Elytral intervals planar, moderately to weakly rugose; integument always black.

Measurements.—Elytral length, 11.9 to 16.6 mm; elytral width, 6.7 to 10.3 mm; pronotal width, 4.9 to 8.4 mm; pronotal length, 4.4 to 6.5 mm.

This series varies considerably, but not according to any obvious geographic pattern. Specimens from the White Mountains and Mono Plateau (fig. 69) tend to be of smaller average size than those from central Nevada and Utah. Owens Valley individuals and a few specimens from Craters of the Moon National Monument, Idaho show definite phenetic similarities to *californica* Mannerheim. Those from south central Utah show reddish black coloration and subreticulate elytral sculpturing suggestive of the Colorado Plateau populations. Specimens from the Kaibab Plateau (fig. 71), are distinctly smaller and more slender than typical Great Basin individuals. This series corresponds to *punctata* LeConte and *tanneri* Blaisdell.

These beetles occupy pinyon-juniper woodland at elevations up to 3,000 m (White Mountains, Inyo County, California), and have also been collected in ponderosa woodland in eastern California and southern Utah. There are a few records from *Artemisia*-dominated basins less than 1,200 m elevation.

Range (fig. 51).—Steens Mountains and Hart Mountains in southeast Oregon, south to the Owens Valley, White Mountains, and Argus Mountains in California and the Charleston Mountains in Nevada, east to Fremont County, Idaho, and the Wasatch Mountains in Utah.

Material examined.—111 ♂ ♂, 122 ♀ ♀.

Colorado Plateau populations (pl. 2, fig. 72).—Elytra intervals weakly sulcate and transversely rugose, producing a reticulate appearance; integument frequently reddish black.

Measurements.—Elytral length, 11.6 to 15.9 mm; elytral width, 6.4 to 9.0 mm; pronotal width, 4.9 to 7.0 mm; pronotal length, 4.4 to 6.5 mm.

As stated above, this series intergrades with the Great Basin populations in southeastern Utah. The peculiar coloration and elytral sculpting is probably an adaptation to the edaphic conditions of the habitat. This series corresponds to *angusta* Casey.

Material examined.—30 ♂ ♂, 27 ♀ ♀ (see fig. 51). Arizona. Coconino County: Tuba City. Navajo County: Adamana; Dun's Place, nr. Navajo Mtn.; Kayenta—18 mi. WNW, 19 mi. SW; 23 mi. W; Marsh Pass; Navajo Mtn.; Winslow.

Colorado. Montezuma County: Cortez.

New Mexico. Bernalillo County: Albuquerque. McKinley County: Gallup. San Juan County: Aztec; San Juan Valley.

Utah. Grand County: Moab. San Juan County: La Sal; La Sal Mtns.; Natural Bridges Nat. Monument; Navajo Mtn.

SUMMARY

This study describes the details of phenetic variation in a genus of tenebrionid beetles (*Coelocnemis*) in western North America. These insects are common inhabitants of conifer and oak-conifer woodland from southern British Columbia to southern Baja California and east to the mountains of Utah and Arizona. Adults are long-lived, nocturnal scavengers. Larvae tunnel through decaying wood or soil.

A number of approaches were employed to illuminate the relationships among these beetles. Gas chromatographic analysis of defensive secretions indicated the presence of two groups of closely related species, one of northern distribution, the other southern, but revealed no significant differences within these species groups.

Results of statistical analyses of geographic variation are based on 81 male and female sample localities from throughout the range of the genus, including a total of about 1,200 specimens. Univariate comparisons of 21 characters were made using Power's modification of Gabriel's Sums of Squares Simultaneous Test Procedure. The two species groups suggested by differences in defensive secretion composition are also apparent on the basis of morphological evidence. The northern series of species (Californica species group) is characterized by coarsely punctate and rugose cuticular surfaces; the southern series (Magna species group) by smooth, finely punctate cuticular surfaces. Within each species group variation is largely gradual and continuous. In the Magna species group a large series of continuously distributed samples in cismontane California is interconnected by a multiple-character cline while the remaining samples are allopatric and differentiated by one or several characters. The most distinct phenotypes represent populations isolated on islands or remote desert mountain ranges. In the Californica species group most of the populations are continuously distributed, but marked geographic variation is apparent in most characters. The members of this species group are homogeneous throughout the extensive Pacific Northwest portion of their range, with most of the variation confined to California populations.

Several multivariate clustering techniques were compared, with results which agreed in most respects but conflicted in many details. The two species groups were segregated by all multivariate methods, and most of the samples were partitioned into analogous clusters. A number of samples, especially the geographically isolated populations of the Magna species group, showed ambiguous relationships which differed among the various clustering methods. In general the multivariate analyses tended to emphasize phenetic gaps between samples, particularly in the case of clines. Among the types of analysis employed, phenograms seemed to represent gradual variation most deceptively. The single nonhierarchical clustering procedure investigated ("taxometric mapping") treated gradual variation more realistically but isolated a large number of samples which showed close relationships to larger clusters according to all other types of analysis. Because of the lack of concordance between clustering techniques and the absence of objective ways of assessing the fit between the results and the data it was concluded that such methods are of limited use for making taxonomic decisions.

The results of the various quantitative analyses were correlated with the available information about paleoclimatic and historical geologic events in western North America and a reconstruction of the probable evolutionary history of *Coelocnemis* is proposed. It is believed that the genus is of northern origin, having adapted to the dry oak and oak-conifer woodland of southwestern United States during Miocene or Pliocene times, when the two species groups probably diverged. Most of the contemporary allopatry probably arose during the Pleistocene, but the populations on Cedros Island and southern Baja California have apparently been isolated since the Pliocene.

A classification recognizing seven species is proposed for the genus. The pronounced character discordance, intermediacy, and gradual nature of much of the variation justify synonymizing most of the previously described species. Reasons were discussed for rejecting the subspecies as a meaningful evolutionary or taxonomic category. Finally a description of the genus and a key and descriptions of the five previously designated and two new species are presented.

APPENDIX 1

List of localities from which data were obtained and analyzed. Information is arranged according to species group and within each group according to species. Asterisks mark samples which were not included in univariate analysis.

Abbreviation	Locality	♂♂	♀♀
	Californica species group		
	punctata		
POWN	Owens Valley, Inyo Co., Calif.	4	16
PWHM	Crooked Creek, White Mts., Inyo Co., Calif.	4	11
PUTH	Various localities, central and southern Utah	12	11
PCHM	Charleston Mts., Clark Co., Nev.	5	4
PNEV	Various localities, central and northern Nevada	7	11
RETC	Various localities, Colorado Plateau, Utah, Ariz., N. Mexico	23	17
	lucia		
CLUC	Santa Lucia Mts., Monterey Co., Calif.	10	8
	californica		
CLAN	San Gabriel Mts., Los Angeles Co., Calif.	9	6
CJAC	San Jacinto Mts., Riverside Co., Calif.	7	8
CSBD	Camp Angeles Maint. Sta., San Bernardino Co., Calif.	7	13
CSCR	Santa Cruz Mts., Sta. Cruz and San Mateo Cos., Calif.	6	7
CALA	Alameda Co., Calif.	6	8
CMAR	Marin Co., Calif.	4	11
CSON	Sonoma Co., Calif.	2*	2*
CNAP	Napa Co., Calif.	5	4
CSMN	Yorkville, Mendocino Co., Calif.	6	7
CNMN	Northern Mendocino Co., Calif.	9	6
CHUM	Weott and Green Point, Humboldt Co., Calif.	5	2*
CTRN	Del Loma, Trinity Co., Calif.	8	7
CSIS	Vicinity of Shasta City, Siskiyou Co., Calif.	10	7
CSHA	Shasta Co., Calif.	6	9
CLAS	Lassen Co., Calif.	7	7
CBUT	Vicinity of Chico, Butte Co., Calif.	7	—*
CPLU	Quincy, Mt. Hough, and Meadow Valley, Plumas Co., Calif.	5	9
CELD	Riverton, El Dorado Co., Calif.	5	10
CMAD	Sugar Pine, Madera Co., Calif.	9	6
CKRN	Alta Sierra, Kern Co., Calif.	8	9
CUTH	Various localities, Utah	9	7
CIDM	Moscow, Idaho and vicinity	13	12
CIDB	Boise, Idaho and vicinity	8	6
CIDC	Craters of the Moon Nat. Mon., Idaho	3*	3*
CORJ	Josephine Co., Oregon	4	3*
CORD	Deschutes and Lake Cos., Oregon	6	7
CORW	Various localities, Willamette Valley, Oregon	4	2*
CORB	Blue Mts., Baker and Union Cos., Oregon	2*	13
CORL	Lake Co., Oregon	4	3*
CORK	Vicinity of Klamath Lake, Klamath Co., Oregon	6	6
CWSH	Various localities, Washington	8	10
CBCO	Oliver, British Columbia	13	7
CBCK	Kamloops and Lillooet, British Columbia	4	5
CBCS	Salmon Arm, British Columbia	7	8
	rugulosa		
RUGU	Various localities, Modoc Plateau, Calif., and Oregon	15	15
	Magna Species Group		
	sulcata		
SPMA	Oracle and vicinity of Tucson, Pima Co., Ariz.	3*	11

| | | Number | |
Abbreviation	Locality	♂♂	♀♀
SPIN	Pinal Mts., Pinal Co., Ariz.	6	8
SYAV	Yarnell and vicinity of Prescott, Yavapai Co., Ariz.	9	6
SNEV	Various localities, central and eastern Nevada	6	7
SYER	Yerrington, Lyon Co., Nev.	7	8
SUTH	Vicinity of St. George, Washington Co., Utah	11	4
SPRO	Cedar Cyn., Providence Mts., San Bernardino Co., Calif.	6	17

slevini

CEDR	Grand Canyon, Cedros Island, Baja California, Mexico	6	5

magna

RCHR	Chiricahua Mts., Cochise Co., Ariz.	9	11
RGIL	Little Green Valley, Gila Co., Ariz.	8	7
RCOC	Williams, Coconino Co., Ariz.	9	10
RCAL	Inyo Co., Calif.	13	12
MSON	Various localities, Sonoma Co., Calif.	8	7
MGLN	Newville, Glenn Co., Calif.	5	1*
MLAK	Upper Lake, Lake Co., Calif.	8	7
MCCO	Contra Costa Co., Calif.	10	5
MSCL	Santa Clara Co., Calif.	9	6
MAMA	Amador Co., Calif.	14	5
MMAD	Madera and Fresno Cos., Calif.	9	5
MVEN	Ventura Co., Calif.	7	8
MSTB	Santa Barbara Co., Calif.	7	8
MSLO	San Luis Obispo Co., Calif.	12	8
MBEN	Pinnacles Nat. Mon., San Benito Co., Calif.	10	10
MMON	Monterey Co., Calif.	6	14
MTUL	Tulare Co., Calif.	3*	12
MNKR	Northern Kern Co., Calif.	7	11
MSKR	Southern Kern Co., Calif.	14	7
MJSH	Joshua Tree Nat. Mon. and vicinity, San Bernardino Co., Calif.	3*	2*
MVIC	Victorville and Yermo, San Bernardino Co., Calif.	5	1*
OANT	Elizabeth Lake and Littlerock, Los Angeles Co., Calif.	6	9
OTOP	Topanga Canyon, Los Angeles Co., Calif.	7	8
ORIV	Vicinity of Palm Springs, Riverside Co., Calif.	7	8
OSDN	Santa Margarita Fire Sta., San Diego Co., Calif.	4	6
OSDG	Vicinity of San Diego, San Diego Co., Calif.	8	7
OSBD	San Bernardino Mts., San Bernardino Co., Calif.	9	11
OBAJ	Various localities, northern Baja California, Mexico	5	2*
MMRT	San Pedro Martir Mts., Baja California, Mexico	3*	3*
CATA	Santa Catalina Is., Los Angeles Co., Calif.	—*	3*

Miscellaneous

CAPE	San Jose del Cabo, Baja California, Mexico	1*	1*
	Total	585	604

LITERATURE CITED

ADAMS, R. P.
    1970. Contour mapping and differential systematics of geographic variation. Syst. Zool. 19: 385–390.

AXELROD, D. I.
    1958a. Evolution of desert vegetation in western North America. Carnegie Inst. Wash. Publ. 590:215–306.
    1958b. Evolution of the Madro-Tertiary Geoflora. Bot. Rev. 24:433–509.
    1966. The Pleistocene Soboba flora of southern California. Univ. Calif. Publ. Geol. Sci. 60:1–79.
    1967. Geologic history of the Californian insular flora, pp. 267–315. In R. N. Philbrick, Proceedings of the Symposium on the biology of the California islands, Santa Barbara Botanical Gardens, Santa Barbara.

AXELROD, D. I., and W. S. TING
    1960. Late Pliocene floras east of the Sierra Nevada. Univ. Calif. Publ. Geol. Sci. 39:1–117.

BAKER, R. G.
    1970. Pollen sequence from late Quaternary sediments in Yellowstone Park. Science 168:1449–1450.

BELTRAN, E., J. S. GARTH and J. M. SAVAGE (eds.)
    1960. The biogeography of Baja California and adjacent seas. Syst. Zool. 9:1–232.

BLAISDELL, F. E.
    1925. Expedition to Guadalupe Island, Mexico, in 1922. Proc. Calif. Acad. Sci. 14:321–343.
    1928. Two new species of *Coelocnemis* (Coleoptera: Tenebrionidae). Pan-Pac. Entomol. 4: 163–165.
    1943. Contributions toward a knowledge of the insect fauna of lower California No. 7 (Coleoptera:Tenebrionidae). Proc. Calif. Acad. Sci. 24:171–288.

BROWN, K. W.
    1971. A population approach to computer taxonomy with applications in the genus *Gonasida* (Coleoptera: Tenebrionidae). Ph.D. thesis, Univ. Calif., Riverside. 317 pp.

CARMICHAEL, J. W., J. A. GEORGE, and R. S. JULIUS
    1968. Finding natural clusters. Syst. Zool. 17:144–150.

CARMICHAEL, J. W., R. S. JULIUS, and P. M. D. MARTIN
    1965. Relative similarities in one dimension. Nature 208: 544–547.

CARMICHAEL, J. W., and P. H. A. SNEATH
    1969. Taxometric maps. Syst. Zool. 18:402–415.

CASEY, T. L.
    1895. Coleopterological Notices 6. Ann. N.Y. Acad. Sci. 8:435–838.
    1924. Memoirs on the Coleoptera, 11:1–347.

CHORLEY, R. J., and P. HAGGETT
    1968. Trend-surface mapping in geographical research, pp. 195–217. In B. J. L. Berry and D. F. Marble, Spatial Analysis, Prentice Hall, N.J.

COOPE, G. R.
    1967. The value of Quaternary insect faunas in the interpretation of ancient ecology and climate, pp. 359–380. In E. J. Cushing and H. E. Wright, Quaternary Paleoecology. Yale Univ. Press, New Haven.
    1970. Interpretation of Quaternary insect fossils. Annu. Rev. Entomol. 14:97–120.

COREY, W. H.
    1954. Tertiary basins of southern California. Calif. Dep. Nat. Resour. Div. Mines Bull. 170: 73–83.

CROVELLO, T. J.
    1968a. The effect of alternation of technique at two stages in a numerical taxonomic study. Univ. Kans. Sci. Bull. 47:761–785.
    1968b. A numerical taxonomic study of the genus *Salix*, section *Sitchensis*. Univ. Calif. Publ. Bot. 44:1–61.

1968c. Key communality cluster analysis as a taxonomic tool. Taxon 17:241–258.

1969. Effects of change of characters and of number of characters in numerical taxonomy. Am. Midl. Nat. 81:68–86.

CUSHING, E. J., and H. E. WRIGHT (eds.)

1967. Quaternary Paleoecology. Proc. VII Congr. Int. Assoc. Quat. Res., Vol. 7. Yale Univ. Press, vii + 433 pp.

DARROW, R. A.

1961. Origin and development of the vegetational communities of the Southwest, pp. 30–47. *In* L. M. Shields and L. J. Gardner, Bioecology of the arid and semiarid lands of the southwest. N. Mex. Highlands Univ. Publ.

DOYEN, J. T.

1970. Biosystematics of the genus *Coelocnemis* (Coleoptera: Tenebrionidae): a quantitative study of variation. Ph.D. thesis, Univ. Calif. Berkeley. 373 pp.

1971. Synopsis of the genus *Oenopion* (Coleoptera: Tenebrionidae: Coelometopini). Coleopt. Bull. 25:109–117.

1972. Familial and subfamilial classification of the Tenebrionoidea (Coleoptera) and a revised generic classification of the Coniontinae (Tentyriidae). Quaest. Entomol. 8:357–376.

DOYEN, J. T., and P. A. OPLER

1973. Distributional affinities of some xerophilous insects (Coleoptera, Lepidoptera) in central California. Southwest. Nat. 18: (in press).

EADES, D. C.

1965. The inappropriateness of the correlation coefficient as a measure of taxonomic resemblance. Syst. Zool. 14:98–100.

1970. Theoretical and procedural aspects of numerical phyletics. Syst. Zool. 19:142–171.

EHRLICH, P. R., and P. H. RAVEN

1969. Differentiation of populations. Science 165:1228–1232.

EISNER. T., F. McHENRY, and M. M. SALPETER

1964. Defensive mechanisms of arthropods. XV. Morphology of the quinone producing glands of a tenebrionid beetle (*Eleodes longicollis* LeC.). J. Morphol. 115:355–400.

EISNER, T., and J. MEINWALD

1966. Defensive secretions of arthropods. Science 153:1341–1350.

FARRIS, J. S.

1969. On the cophenetic correlation coefficient. Syst. Zool. 18:279–285.

FARRIS, J. S., A. G. KLUGE and M. J. ECKARDT

1970. A numerical approach to phylogenetic systematics. Syst. Zool. 19:172–189.

FINDLEY, J. S., and S. ANDERSON

1956. Zoogeography of the montane mammals of Colorado. J. Mammal. 37:80–82.

FIORI, G.

1954a. Sulla formazione de spermatofori in un Coelottero Tenebrionide della Tripolitania. R. C. Accad. Lincei (Rome) 8:440–444.

1954b. Morphologia addominale, anatomia et istologia degli apparati genitali di *Pimelia angulata confalonierii* Grid. (Coleoptera Tenebrionidae) e formazione dello spermatoforo. Boll. Ist. Entomol. Univ. Studi Bologna 20:377–422.

FISHER, D. R., and F. J. ROHLF

1969. Robustness of numerical taxonomic methods and errors in homology. Syst. Zool. 19:33–36.

FLAKE, R. H., and B. L. TURNER

1968. Numerical classification for taxonomic problems. J. Theor. Biol. 20:260–270.

FLINT, R. F.

1957. Glacial and Pleistocene Geology. Wiley, New York. 533 pp.

GABRIEL, K. R.

1964. A procedure for testing the homogeneity of all sets of means in analysis of variance. Biometrics 20:459–477.

GABRIEL, K. R., and R. R. SOKAL

1969. A new statistical approach to geographic variation analysis. Syst. Zool. 18:259–278.

GISSLER, C. G.

1879a. Biological notes on some genera of Tenebrionidae. Bull. Brooklyn Entomol. Soc. 2:7–8.

1879b. On the repugnatorial glands in *Eleodes*. Psyche 2:209–210

GLEASON, H. A.

1926. The individualistic concept of the plant association. Torrey Bot. Club Bull. 53:7–26.

GOWER, J. C.

1967. A comparison of some methods of cluster analysis. Biometrics 23:623–637.

HALL, A. V.

1969. Avoiding informational distortion in automatic grouping programs. Syst. Zool. 18: 318–329.

HAPP, G. M.

1968. Quinone and hydrocarbon production in the defensive glands of *Eleodes longicollis* and *Tribolium castaneum* (Coleoptera, Tenebrionidae). J. Insect Physiol. 14:1821–1837.

HARRY, O. G.

1969. Effect of gregarines on *Tenebrio molitor*. Tribolium Inform. Bull. 11:81–82.

HAYASHI, N.

1966. A contribution to the knowledge of the larvae of Tenebrionidae occurring in Japan. Insecta Matsumurana, Suppl. 1:1–41.

1968. Additional notes on the larvae of Lagriidae and Tenebrionidae occurring in Japan. Insecta Matsumurana, Suppl. 3:1–12.

HEUSSER, C. J.

1960. Late Pleistocene environments of North Pacific North America. Am. Geogr. Soc. Spec. Publ. 35:1–308.

HOLLOWAY, J. D., and N. JARDINE

1968. Two approaches to zoogeography: a study based on the distribution of butterflies, birds and bats in the Indo-Australian area. Proc. Linn. Soc. Lond. 179:153–188.

HORN, G. H.

1871. Revision of the Tenebrionidae of America North of Mexico. Trans. Am. Philos. Soc. 14:253–404.

HOWDEN, H. F.

1963. Speculations on some beetles, barriers, and climates during the Pleistocene and pre-Pleistocene periods in some non-glaciated portions of North America. Syst. Zool. 12: 178–201.

1966. Some possible effects of the Pleistocene on the distributions of North American Scarabaeidae (Coleoptera). Can. Entomol. 98:1178–1190.

HUBAC, J. M.

1964. Application de la taxonomie de Wraclaw (technique des dendrites) à quelques populations du *Campanula rotundifolia* L., s.l., et utilisation de cette technique pour l'etablissement des clés de détermination. Soc. Bot. France 111:331–346.

HUBBS, C. L. (ed.)

1958. Zoogeography. Am. Assoc. Adv. Sci., Publ. 51: 1–509.

HULL, D. L.

1970. Contemporary systematic philosophies. Annu. Rev. Syst. Ecol. 1:19–54.

JAMES, F. C.

1970. Geographic size variation in birds and its relationship to climate. Ecology 51:365–390.

JARDINE, N., and R. SIBSON

1971. Mathematical Taxonomy. John Wiley & Sons Ltd., London, xviii + 286 pp.

JOHNSON, A. W.

1968. The evolution of desert vegetation in western North America, pp. 101–140. *In* G. W. Brown, (ed.), Desert Biology. Academic Press, N. Y.

JOHNSTON, R. F.

1969. Character variation and adaptation in European sparrows. Syst. Zool. 18:206–231.

JOHNSTON, R. F., and R. K. SELANDER
1964. House sparrows: rapid evolution of races in North America. Science 144:548–550.
JOLICOEUR, P.
1959. Multivariate geographical variation in the wolf *Canis lupis* L. Evolution 13:283–299.
KAUFMANN, T.
1966. Observations on some factors which influence aggregation by *Blaps sulcata* Sol. (Coleoptera: Tenebrionidae) in Israel. Ann. Entomol. Soc. Am. 59:660–664.
KENDALL, D. A.
1968. The structure of the defense glands in Alleculidae and Lagriidae. Trans. R. Entomol. Soc. Lond. 120:139–156.
KING, P. B.
1958. Evolution of modern surface features of western North America, pp. 3–60. *In* C. L. Hubbs, (ed.) Zoogeography. Am. Assoc. Adv. Sci. Publ. 51.
KOCH, C.
1960. Monograph of the Tenebrionidae of southern Africa, Vol. I. Transvaal Mus. Mem. 7: 1–242.
KRUSKAL, J. B.
1964. Multidimensional scaling by optimizing goodness of fit to a nonmetric hypothesis. Psychometrika 29:1-27.
LACORDAIRE, T.
1859. Histoire naturelle des Insects. Genera des Coléoptères 5:1–750.
LADISCH, R. K.
1967. Quinoid secretions in grain and flour beetles. Nature 215:939–940.
LARSON, R. L., H. W. MENARD and S. M. SMITH
1968. Gulf of California: a result of ocean-floor spreading and transform faulting. Science 161:781–784.
LECONTE, J. L.
1851. Descriptions of new species of Coleoptera from California. Ann. Lyceum Nat. Hist. N.Y. 5:125–216.
1854. Descriptions of new Coleoptera collected by Thos. H. Webb, M.D., in the years 1850–'51 and '52 while secretary to the U. S. and Mexican Boundary Commission. Proc. Acad. Nat. Sci. Phila. 7 (1855):220–225.
LECONTE, J. L., and G. H. HORN
1883. Classification of the Coleoptera of North America. Smithson. Misc. Coll. 26:1–567.
LIDICKER, W. Z.
1960. An analysis of intraspecific variation in the kangaroo rat *Dipodomys merriami*. Univ. Calif. Publ. Zool. 67:125–218.
LILES, M. P.
1956. A study of the life history of the forked fungus beetle, *Bolitotherus cornutus* Panzer (Coleoptera: Tenebrionidae). Ohio J. Sci. 56:329–337.
LINNELL, M.
1901. Descriptions of some new species of North American Heteromerous Coleoptera. Proc. Entomol. Soc. Wash. 4:180–188.
LINSLEY, E. G.
1958. Geographical origins and phylogenetic affinities of the Cerambycid beetle fauna of western North America, pp. 299–320. *In* C. L. Hubbs, Zoogeography. Am. Assoc. Adv. Sci. Publ. 51.
LINSLEY, E. G., and J. A. CHEMSAK
1961. A distributional and taxonomic study of the genus *Crossidius*. Misc. Publ. Entomol. Soc. Am. 3:25–64.
MANNERHEIM, C. G.
1843. Beitrag zur Kaefer-fauna der Aleutischen Inseln, der Insel Sitkha und Neu-Californiens. Bull. Soc. Imp. Nat. Moscou 16:175–314.
1844. Genus *Coelocnemis* Mannerheim. Mag. Zool. 6 (ser. 2): pt. 133, pp. 1–4.

MASON, H. L.
  1947. Evolution of certain floristic associations in western North America. Ecol. Monogr.
    17:203–210.
MARCUS, L. F., and J. H. VANDERMEER
  1966. Regional trends in geographic variation. Syst. Zool. 15:1–13.
MARTIN, P. S., and P. J. MEHRENGER
  1965. Pleistocene pollen analysis and biogeography of the Southwest, pp. 433–451. *In* H. E.
    Wright and D. G. Frey, The Quaternary of the United States. Princeton Univ. Press,
    Princeton.
MEHRENGER, P. J.
  1965. Late Pleistocene vegetation in the Mojave Desert of southern Nevada. J. Ariz. Acad.
    Sci. 2:172–188.
MICHENER, C. D.
  1970. Diverse approaches to systematics, pp. 1–38. *In* T. Dobzhansky, M. K. Hecht and W. C.
    Steere, Evolutionary Biology, Vol. 4. Appleton-Century-Crofts, N.Y.
MICHENER, C. D., and R. R. SOKAL
  1966. Two tests of hypotheses of nonspecificity in the *Hoplitis* complex. Ann. Entomol. Soc.
    Am. 59:1211–1217.
MINKOFF, E. C.
  1965. The effects on classifications of slight alterations in numerical technique. Syst. Zool.
    14:196–213.
MOORE, B. P., and B. E. WALLBANK
  1968. Chemical compositon of the defensive secretion in carabid beetles and its importance
    as a taxonomic character. Proc. R. Entomol. Soc. Lond. (B) 37:62–72.
MOSS, W. W.
  1967. Some new analytic and graphic approaches to numerical taxonomy, with an example
    from the Dermanyssidae (Acari). Syst. Zool. 16:177–207.
  1968. Experiments with various techniques of numerical taxonomy. Syst. Zool. 17:31–47.
MOSS, W. W., and J. A. HENDRICKSON
  1973. Numerical taxonomy. Annu. Rev. Entomol., 18:227–258.
MOSS, W. W., and W. A. WEBSTER
  1969. A numerical taxonomic study of a group of selected strongylates (Nematoda). Syst.
    Zool. 18:423–443.
OXNARD, E. E., and P. M. NEELY
  1969. The descriptive use of neighborhood limited classification in functional morphology:
    an analysis of the shoulder in primates. J. Morphol. 129:127–148.
PACE, A. E.
  1967. Life history and behavior of a fungus beetle, *Bolitotherus cornutus* (Tenebrionidae).
    Univ. Mich. Mus. Zool. Occas. Pap. 653:1–15.
PAPP, C. S.
  1961. Checklist of the Tenebrionidae of America, north of the Panama Canal. (Notes on
    North American Coleoptera, No. 14). Opusc. Entomol. 26:97–140.
PAUKEN, R. J., and D. E. METTER
  1971. Geographic representation of morphologic variation among populations of *Ascaphus
    truei* Stejneger. Syst. Zool. 20:434–441.
PEABODY, F. E., and J. M. SAVAGE
  1958. Evolution of a coast range corridor in California and its effect on the origin and
    dispersal of living amphibians and reptiles, pp. 159–186. *In* C. L. Hubbs, Zoogeography.
    Am. Assoc. Adv. Sci. Publ. 51.
PHILBRICK, R. N.
  1967. Proceedings of the Symposium on the Biology of the California Islands. Santa Barbara
    Botanical Gardens, Santa Barbara. 363 pp.
POWER, D. M.
  1970. Geographic variation of red-winged blackbirds in central North America. Univ. Kans.
    Mus. Nat. Hist. Publ. 19:1–83.

PRIM, R. C.
  1957. Shortest connection networks and some generalizations. Bell Syst. Tech. J. 36:1389–1401.
RAY, C.
  1960. The application of Bergmann's and Allen's rules to poikilotherms. J. Morphol. 106: 85–108.
RISING, J. D.
  1970. Morphological variation and evolution in some North American orioles. Syst. Zool. 19:315–351.
ROHLF, F. J.
  1967. Correlated characters in numerical taxonomy. Syst. Zool. 16:109–126.
  1968. Stereograms in numerical taxonomy. Syst. Zool. 17:246–255.
  1970. Adaptive hierarchical clustering schemes. Syst. Zool. 19:58–82.
ROHLF, F. J., and D. R. FISHER
  1968. Tests for hierarchical structure in random data sets. Syst. Zool. 17:407–412.
ROTH, L. M., and T. EISNER
  1962. Chemical defenses of arthropods. Annu. Rev. Entomol. 7:107–136.
ST. GEORGE, R. A.
  1924. Studies on the larvae of North American beetles of the subfamily Tenebrioninae with a description of the larva and pupa of *Merinus laevis* Olivier. Proc. U. S. Natl. Mus. 65:1–22.
SCHILDKNECHT, H., K. HOLOUBEK, and H. KRAMER
  1964. Defensive substances of arthropods, their isolation and identification. Angew. Chem. 3:73–82.
SHARSMITH, H. K.
  1945. Flora of the Mount Hamilton Range of California. Am. Midl. Nat. 34:287–367.
SNEATH, P. H. A.
  1967. Trend-surface analysis of transformation grids. J. Zool. (Lond.) 151:65–122.
  1969. Recent trends in numerical taxonomy. Taxon 18:14–20.
SNEATH, P. H. A., and R. R. SOKAL
  1973. Numerical Taxonomy. W. H. Freeman and Co., San Francisco (in press).
SOKAL, R. R.
  1969. Biometrical techniques in systematics (Discussion), pp. 589–592. *In* C. G. Sibley (ed.), Systematic Biology. Natl. Acad. Sci. Publ. 1692.
SOKAL, R. R., and T. J. CROVELLO
  1970. The biological species concept: a critical evaluation. Am. Nat. 104:127–153.
SOKAL, R. R., and C. D. MICHENER
  1967. The effects of different numerical techniques on the phenetic classification of bees of the *Hoplitis* complex (Megachilidae). Proc. Linn. Soc. Lond. 178: 59–74.
SOKAL, R. R., and P. H. A. SNEATH
  1963. Principles of Numerical Taxonomy. W. H. Freeman and Co., San Francisco. xvi + 359 pp.
STEBBINS, G. L., and J. MAJOR
  1965. Endemism and speciation in the California flora. Ecol. Monogr. 35:1–35.
STEBBINS, R. C.
  1949. Speciation in salamanders of the plethodontid genus *Ensatina*. Univ. Calif. Publ. Zool. 48:377–526.
  1954. Amphibians and Reptiles of Western North America. McGraw-Hill, New York, xii + 536 pp.
  1957. Intraspecific sympatry in the lungless salamander *Ensatina eschscholtzi*. Evolution 11: 265–270.
  1958. A new alligator lizard from the Panamint Mountains, Inyo County, California. Am. Mus. Novit. 1883:1–27.
THEODORIDES, J.
  1952. Contributions a l'étude écologique des parasites et commensaux de Coléoptères (2ᵉ note) Trans. 9th Int. Congr. Entomol. Amsterdam, 1951 1:454–459.

1966. Sur la repartition geographique de *Stylocephalus filiformis* Theod. 1959 (*Eugregarina stylocephalidae*). Vie Milieu 17 (2c):843–844.

THEODORIDES, J., and P. JOLIVET
1964. Gregarines parasites des Coléoptères D'Ethiopie. Ann. Parasitol. Hum. Comp. 39:1–31.

TSCHINKEL, W. R.
1968. Studies on the sex pheromones and defensive secretions of tenebrionid beetles. Ph.D. thesis, Univ. Calif., Berkeley. iii + 34 pp.
1972. 6-alkyl-1, 4-naphthoquinones from the defensive secretion of the tenebrionid beetle, *Argoporis alutacea*. J. Insect Physiol. 18:711–722.

VALENTINE, J. W., and J. H. LIPPS
1967. Late Cenozoic history of the southern California Islands, pp. 21–35. *In* R. N. Philbrick, Proceedings of the symposium on the biology of the California Islands. Santa Barbara Botanical Gardens. Santa Barbara.

VAN DYKE, E. C.
1919. The distribution of insects in western North America. Ann. Entomol. Soc. Am. 12:1–12.
1939. The origin and distribution of the Coleopterous insect fauna of North America. Proc. 6th Pac. Sci. Congr. 4:255–268.

WADE, J. S.
1921. Biology of *Embaphion muricatum*. J. Agric. Res. 22:323–334.

WADE, J. S., and R. A. ST. GEORGE
1923. Biology of the false wire worm *Eleodes suturalis* Say. J. Agric. Res. 26:547–566.

WELLS, P. V.
1970. Postglacial vegetational history of the Great Plains. Science 167:1574–1582.

WELLS, P. V., and R. BERGER
1967. Late Pleistocene history of coniferous woodland in the Mohave Desert. Science 155:1640–1647.

WELLS, P. V., and C. D. JORGENSEN
1964. Pleistocene wood rat middens and climate change in Mohave Desert: a record of juniper woodlands. Science 143:1171–1173.

WHITTAKER, R. H.
1965. Dominance and diversity in land plant communities. Science 147:250–260.

WHITTAKER, R. H., and P. P. FEENY
1971. Allelochemics: chemical interactions between species. Science 171:755–770.

WHITTAKER, R. H., and W. A. NIERING
1968. Vegetation of the Santa Catalina Mountains, Arizona. III. Species distribution and floristic relations on the North Slope. J. Ariz. Acad. Sci. 5:3–21.

WOLFE, J. A.
1969. Neogene floristic and vegetational history of the Pacific Northwest. Madrono 20:83–110.

WRIGHT, H. E., and D. G. FREY
1965. The Quaternary of the United States. Princeton Univ. Press, Princeton, 922 pp.

PLATES

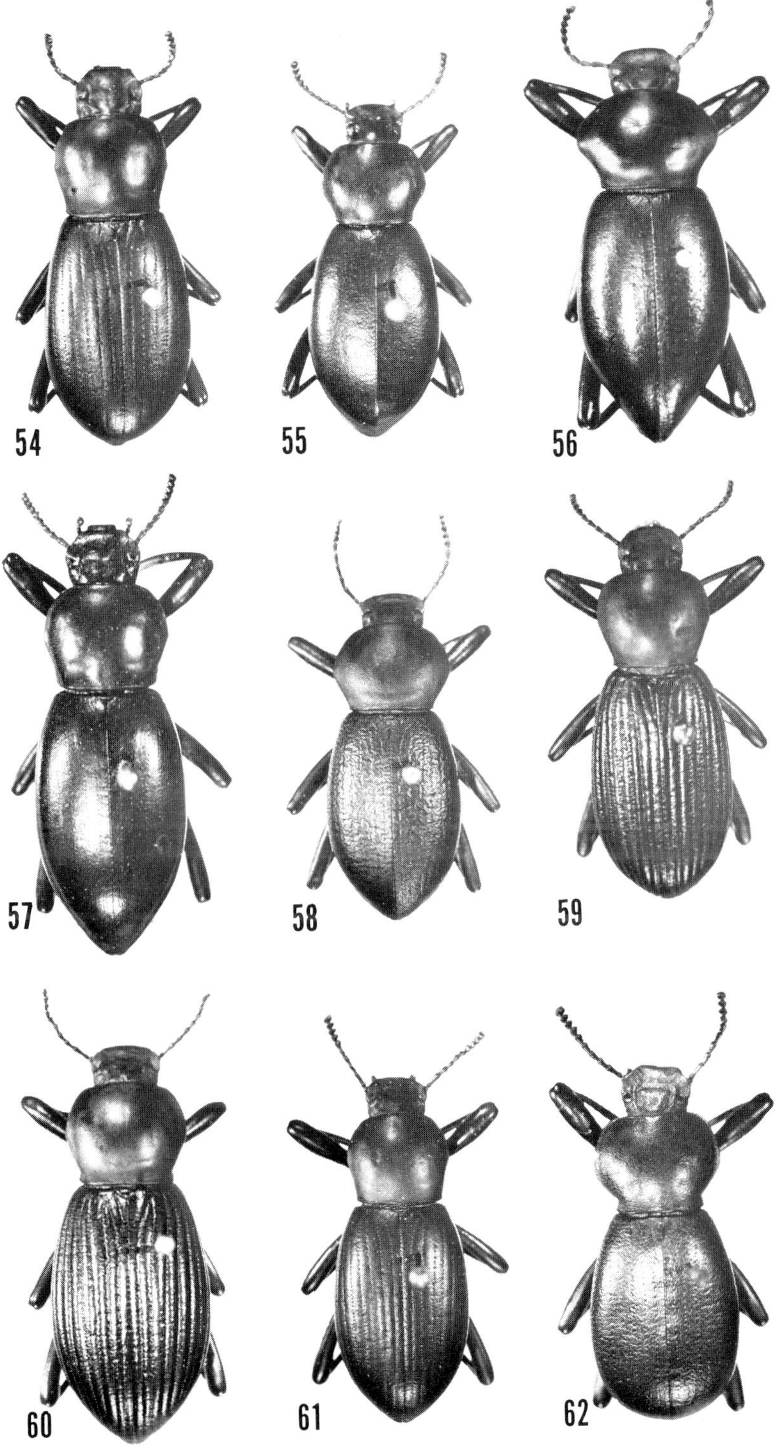

PLATE 2

(Figs. 63–71)

63. *Coelocnemis californica* Mannerheim; Sierra Nevadan phenotype, female
(California, Madera County, Ahwannee).

64. *C. californica;* southern California phenotype, female (California, San Bernardino County,
0.4 mi. NE Camp Angeles Maintenance Station [San Bernardino Mountains]).

65. *C. californica;* San Francisco Bay region phenotype, female (California,
San Mateo County, Pescadero).

66. *C. slevini* Blaisdell, female (Mexico, Baja California, Cedros Island).

67. *C. lucia* n. sp., female (California, Monterey County, Nacimiento Summit, 5 air mi. E Lucia).

68. *C. rugulosa* n. sp., female (California, Siskiyou County, Shasta Springs).

69. *C. punctata* LeConte, female (California, Inyo County, Whitney Portal). Specimens from
this locality and others along the eastern escarpment of the Sierra Nevada Range are intermediate between typical *punctata* and *californica,* especially in prothoracic dimensions (see
figs. 13, 15, 17).

70. *C. punctata;* Great Basin phenotype, female (Utah, Beaver County, Beaver Valley).

71. *C. punctata;* Great Basin phenotype, female (Arizona, Coconino County, Bright Angel Camp,
Grand Canyon National Park). Note the reduced punctation and narrow form, as compared
to figure 70.

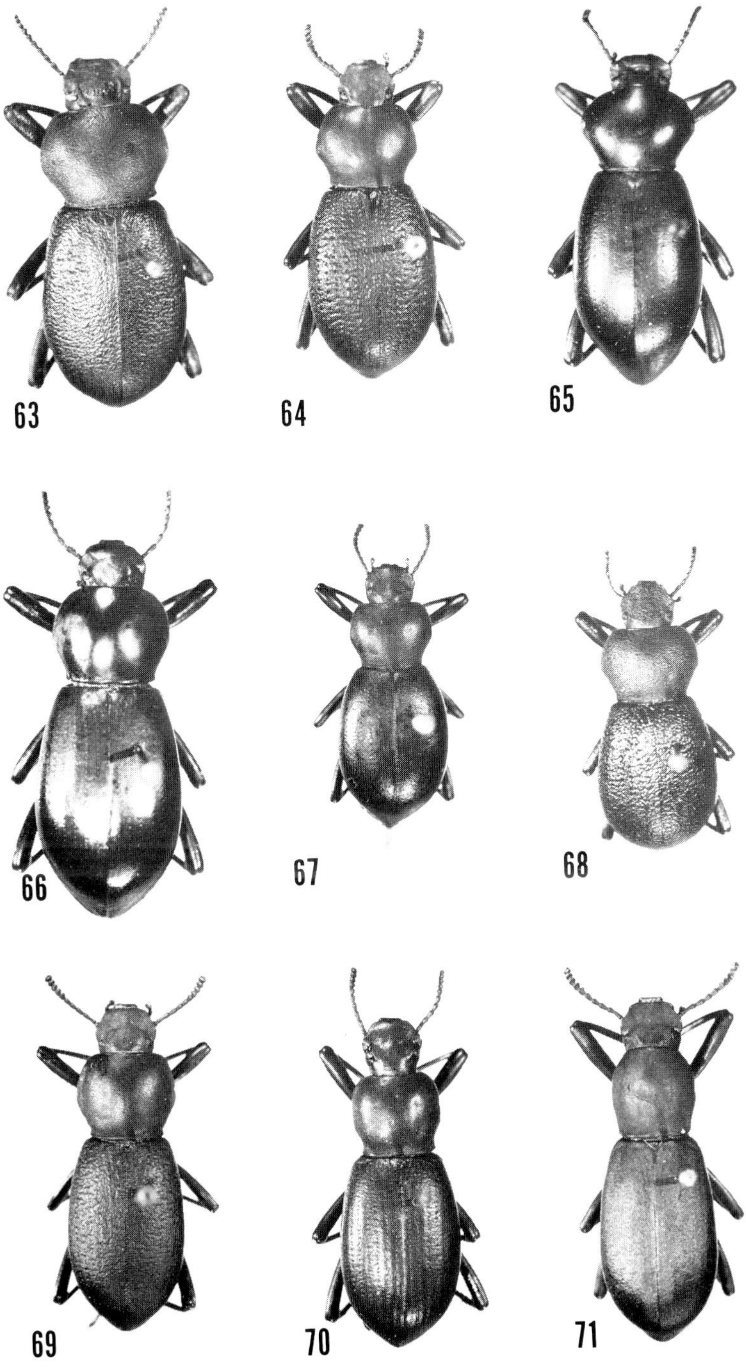
63
64
65
66
67
68
69
70
71

PLATE 3

(Figs. 72–74)

72. *Coelocnemis punctata* LeConte; Colorado Plateau phenotype
(Arizona, Navajo County, Winslow).
73. *C. californica* Mannerheim, dorsal aspect of pronotum and head,
showing characteristic sculpturing.
74. *C. rugulosa* n. sp., dorsal aspect of pronotum and head, showing coarse, reticulate sculpturing.

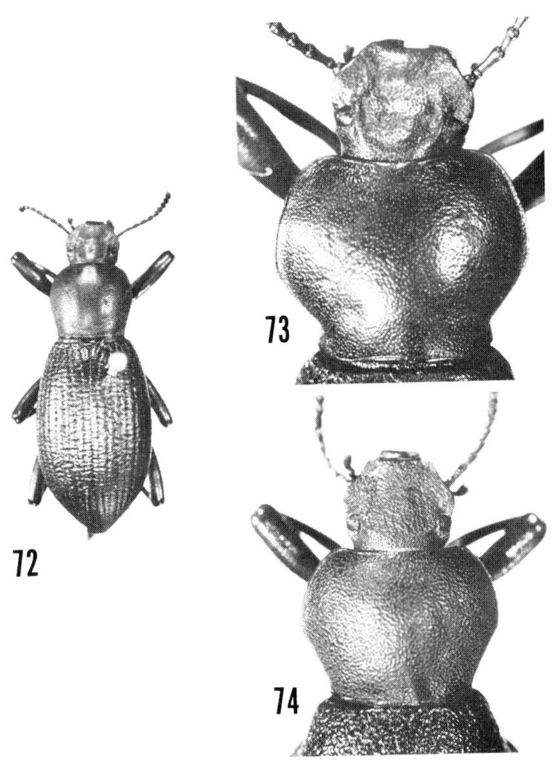